# UNIVERSE

From 13.8 billion years ago to the infinite future

COVER IMAGE: The collision of the galaxies NGC 4038 and NGC 4039 has disrupted their spiral structures and prompted the formation of a vast number of stars. This image was taken by the Hubble Space Telescope. *(NASA/ESA/STScI/AURA/Brad Whitmore)*

## Acknowledgements

I would like to thank, in no particular order, Steve Rendle, W. David Woods, Ford Renton, and James Joel Knapper.

First published in March 2019

A catalogue record for this book is available from the British Library.

ISBN 978 1 78521 209 3

Library of Congress control no. 2018938914

Published by Haynes Publishing,
Sparkford, Yeovil,
Somerset BA22 7JJ, UK.
Tel: 01963 440635
Int. tel: +44 1963 440635
Website: www.haynes.com

Haynes North America Inc.,
859 Lawrence Drive, Newbury Park,
California 91320, USA.

Printed in Malaysia.

# UNIVERSE

**From 13.8 billion years ago to the infinite future**

## Owners' Workshop Manual

An insight into the study of the universe and our place in it

**David M. Harland**

# Contents

**OPPOSITE** The 100in Hooker reflecting telescope on Mount Wilson in California that enabled Edwin Hubble to explore the universe beyond our own galaxy.

*(Creative Commons/Ken Spencer)*

# Introduction

It used to be thought that the universe consisted of Earth, its Moon, the Sun, the planets, and the fixed stars. We now know it to be much bigger than that.

What we see as the Milky Way glowing in the night sky was discovered to be a lens-shaped structure of stars. The Sun is merely a star that appears bright to us because it is so close. There are other systems of stars similar to the Milky Way system. We call them galaxies and they are generally flying away from one another because the universe is in a state of expansion. As physicists investigated the origin of the universe, they called the creation event the Big Bang.

It is remarkable that just a few numbers, the physical constants of nature, define the key features of the cosmos. If the value of any one of these numbers had been significantly different, this could have made it impossible for life to develop. If a 'multiverse' of parallel universes exists in which the values of these numbers are different, then our very existence can be attributed to the fact that this particular universe is configured for life. Indeed, it might even be argued that the purpose of our universe was to give rise to life.

There are supermassive 'black holes' in the cores of galaxies. We can detect a black hole by the radiation emitted by the material it draws in. When the viewing geometry is favourable, we see a glowing quasar. When the black hole has sucked in all of the nearby material, it ceases to emit intense ionising radiation, removing the threat to the development of life in the host galaxy.

Earth is currently the only place that we know to host life, but we are seeking evidence of its existence elsewhere. Historically, the favourite candidate has been Mars. We have robots on the surface studying environments that might have been conducive to life in ancient times when the planet was warmer and wetter than it is today. We also have sensors orbiting Mars, 'sniffing' for chemicals that should determine whether microbial life exists in environments isolated from the hostile surface. In addition, we have found oceans beneath the icy crusts of bodies in the outer solar system and it is possible that hydrothermal vents there might support life, just as they do on the ocean floors of Earth.

We have gone from wondering whether other stars have systems of planets to finding that they are common. The configuration of our own solar system has proved far from typical, and the variety is challenging theorists. The search is on for a star that has an Earth-like planet with an atmosphere whose composition is 'out of equilibrium' in a manner suggestive of life.

If it turns out that life has originated independently on different bodies in our solar system, and also in other such systems, then we might reasonably infer that it is likely to be ubiquitous in the universe.

It is conceivable that we are the first intelligent species to develop in the history of a universe that is almost 14 billion years old, but the statistics suggest otherwise. There are hundreds of billions of stars in our galaxy, a fair proportion of them similar to the Sun, and the universe contains an even greater number of galaxies.

A methodology has been conceived to estimate the number of potentially communicative extra-terrestrial civilisations in our galaxy, but we have no way of arriving at a result. Attempts to detect radio communications have thus far been unsuccessful. The Fermi paradox contrasts the high probability of the existence of extra-terrestrial civilisations with the lack of evidence for them.

Our study of the universe therefore covers a broad range of topics, spanning the size range from subatomic particles to structures on the largest possible scale, and mysteries that include the nature of 'dark matter' and 'dark energy', and the origin and distribution of life.

So let's get started...

**OPPOSITE Radio dishes of the Atacama Large Millimetre/ submillimetre Array (ALMA) in the Atacama Desert of Chile, with the central region of Milky Way above.** *(ESO/B. Tafreshi)*

## Chapter One

# Awakening

The first step in our scientific study of the universe was the discovery that universal gravitation explains how celestial bodies move. Next it was found that the Sun is just another star, and that the stars are distributed in space in a lens-shaped structure of enormous size. Far from constituting almost the entire universe, the solar system is actually tiny.

**OPPOSITE** A depiction of an inquiring mind probing beyond the classical dogma of a firmament of stars. It was published as a woodcut by Camille Flammarion in his 1888 volume *L'atmosphère: Météorologie Populaire* and then colourised in 1998 by Heikenwaelder Hugo.

# Universal gravitation

Isaac Newton was born in the village of Woolsthorpe in Lincolnshire several months after his father, a farm hand, had passed away. When his mother Hannah remarried, Newton was sent to live with his grandmother in Grantham, where he started his schooling. Aged 16, he was called back to the farm to help his mother following the death of his stepfather, but he was of little assistance because his mind was distracted by intellectual pursuits.

In 1661 Newton was admitted to Trinity College, Cambridge. On graduating in 1665 he remained for an advanced degree. Isaac Barrow, the Lucasian professor of mathematics, recognising Newton's exceptional mathematical talent, urged him to study the nature of light. This had been considered in a philosophical manner, with little experimentation, and opinion differed regarding whether light resembled a liquid or a vapour. Once Newton had read all the available literature he set about devising and conducting his own experiments.

A plague broke out in London in August 1665 and when it started to migrate north, the university sent its students home. The self-motivated Newton spent the next two years in enforced isolation pursuing his investigations and, if anything, he benefited from the absence of distractions. As he broke ground in understanding not only the nature of light but also how objects moved and interacted with each other, he developed new mathematical formalisms to describe his discoveries.

In particular, Newton noticed that if he darkened his laboratory and allowed a single ray of light to strike a glass prism, a coloured band would be projected on the wall beyond. He wasn't the first to observe the rainbow produced by a prism, but he was the first to realise why it did so. At that time, sunlight was believed to be 'pure' white light and that colours were 'impurities'. But Newton realised that white light did not indicate the absence of

**BELOW A portrayal of Isaac Newton by Jean-Leon Huens, featuring the key elements of his scientific work.** *(National Geographic Creative/Alamy Stock Photo)*

colour; it was a blend of colours that ranged
from violet to red, which the prism 'refracted' to
differing degrees, being greatest in the case of
violet. He proved this discovery by inserting an
inverted prism into the rainbow as it emerged
from a regular prism and thereby reformed the
ray of white light. He decided to call the artificial
rainbow a 'spectrum'.

Noting that light didn't possess mass,
Newton concluded that light comprised tiny
massless 'corpuscles'. However, he could not
explain why a prism refracted corpuscles of
different colours to differing degrees. When
light was refracted upon passing through a
lens, the resulting image was marred by glaring
colours. Having noted that a mirror was free
of this 'chromatic aberration', in 1667 Newton
developed a telescope in which a curved mirror
focused the light.

The Royal Society of London had been
established in 1660 in order to debate new
scientific ideas. After Newton showed one
of his telescopes to the Society in 1671, the
secretary, Henry Oldenburg, was so impressed
that he asked Newton to become a member of
the Society. Newton was reluctant, but in 1669
Barrow had relinquished his chair to his pupil
and so, as a Cambridge don, Newton accepted
and promised to keep the Society appraised
of his work. Accordingly, in February 1672 he

submitted a paper entitled 'New Theory about
Light and Colour' in which he summarised what
he had learnt from the dozens of experiments
he had undertaken while in isolation during
the plague. Robert Hooke, the curator of the
Society, who in 1665 had published an account
of his own studies into light, took exception.
At Oldenburg's behest, Newton engaged with
Hooke in a debate-by-letter that soon expanded
as other people offered their opinions. After
three years, Newton, resenting the time that this
consumed, not only ceased to respond but also
decided that he would not communicate future
investigations to the Society.

In 1666, wondering why an apple should
fall straight down and yet the Moon revolves
around Earth, Newton realised that they must
both feel the same force. In developing a
formalism, he realised that the strength of
the force between any two objects is directly
proportional to the product of their masses
and inversely so to the square of the distance
separating them, measured centre to centre.
Since his equations required it, Newton
introduced the concept of 'action at a distance'
as the means by which gravity acted, but he
refused to speculate about how it came about.

While an undergraduate at Oxford,
Edmund Halley wrote a book in which
he discussed the laws of orbital motion

developed by Johannes Kepler earlier in the century; primarily that the planets travel around the Sun in elliptical orbits.

In 1676, John Flamsteed, the Astronomer Royal, read Halley's book, hired him, and dispatched him with a 24ft telescope to the island of St Helena in the South Atlantic to set up the first observatory in the southern hemisphere, which was essentially virgin territory for astronomers. Despite poor weather, Halley was able to compile a catalogue of 341 southern stars. On his return to Oxford in 1678 he was awarded an advanced degree for this work and elected to the Royal Society.

When Hooke told Halley in 1684 that he had discovered that gravity obeyed an 'inverse square' law, he was invited to provide his mathematical proof of this relationship. After stalling, Hooke said that he could not actually prove it. Shortly thereafter, while paying Newton a visit, Halley asked whether Newton had given consideration to the force that made the planets orbit the Sun. Newton said that he had calculated it five years previously. He was unable to find his notes amongst the piles of papers but promised to rework the proof and forward it. When Halley saw the proof, he was so impressed by the elegance of the method that he urged Newton to write up in book form everything that he had discovered of the laws of motion.

In 1670 Newton had invented a mathematical method that he called 'fluxions' (which we call 'calculus') and had used it in his study of motion, but for the book he laboriously proved all of his theorems in traditional geometric style in order to help others to follow his reasoning. As it was a scholarly work, he wrote in Latin. The monumental *Philosophiae Naturalis Principia Mathematica* was published in 1687.

After using Newton's law of gravitation to calculate the orbits of historical comets, Halley suspected in 1705 that several viewings with essentially the same highly elliptical orbit were repeated appearances of a single comet. He predicted that it would reappear in 1758, which it did, and although he died in 1742 and so was unable to witness the return of the comet, it has entered the history books as Halley's Comet. This triumphant prediction banished any lingering doubt as to the validity of Newton's law of gravitation.

## A system of stars

In 1710 Halley gained an interest in ancient star catalogues, particularly the one by Hipparchus, the greatest of the classical astronomers, in the 2nd century BC. It had long been known that the positions of the stars drifted in a systematic manner. Halley realised that this meant Earth's axis of rotation was 'precessing' with a period of about 25,000 years, causing each of the celestial poles to trace out a circle with a radius matching the angle of tilt.

What is more, in 1718 Halley noted that the bright stars Sirius, Aldebaran and Arcturus all showed 'proper motions'. Sirius was travelling across the sky at a rate of 1.3 seconds of arc per annum. It had travelled a distance equivalent to one and a half times the angular diameter of the Moon since classical times.

The discovery that stars pursue individual motions across the sky marked the downfall of the concept of the 'celestial sphere' as a fixed realm. And the fact that some stars moved more than others implied that they were at different distances. Furthermore, because the stars which were seen to be moving were amongst the brightest, this suggested they were the closest to us.

**BELOW** The title page of the 1726 edition of Isaac Newton's *Principia*.

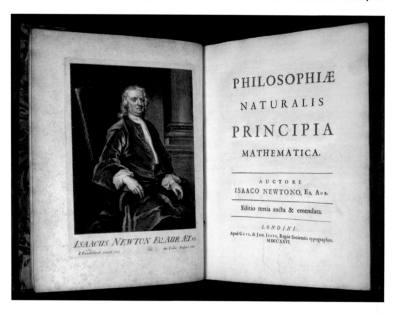

PHILOSOPHIÆ

NATURALIS

PRINCIPIA

MATHEMATICA.

AUCTORE
ISAACO NEWTONO, Eq. Aur.

Editio tertia aucta & emendata.

LONDINI:
Apud Guil. & Joh. Innys, Regiæ Societatis typographos.
MDCCXXVI.

ISAACUS NEWTON Eq.AUR.ÆT 83.

Parallax is the displacement in the apparent position of an object when it is viewed on two different lines of sight.

The parallax of a star is one half of the angle between lines of sight, six months apart, when Earth is on opposite sides of its orbit of the Sun with the diameter of the orbit perpendicular to the direction of the star.

**ABOVE A depiction of Tycho Brahe in his observatory, pointing to a slot in the wall through which an assistant sights the large quadrant. It is a hand-coloured engraving from his *Astronomiæ Instauratæ Mechanica*, published in 1598.**

Unable to detect the 'parallax' of Sirius across the baseline of the diameter of Earth's orbit around the Sun, Halley reasoned that if the star was as luminous as the Sun then it must be 120,000 times more distant in order to appear as faint as it does. At this distance its parallax would be 1 second of arc, which was 200 times finer than was attained by the large quadrant used by Tycho Brahe to measure the positions of the stars in the late 16th century, and too small an angle for Halley to measure using his own telescope.

James Bradley set out in 1725 to measure the parallax of Gamma Draconis. He chose this star because it was bright and because, from London, it passed almost directly overhead throughout the year. However, after taking various sources of error into account in 20 years of observations, he did not find any measurable parallax. Believing he could measure a position to 1 second of arc, he could only conclude that the star was very far away. If such a bright star was so far off, how much farther away must the fainter stars be from us?

Democritus of Thrace speculated in 425 BC that the band of light that spans the celestial realm as the Milky Way consisted of an uncountable number of stars, but this idea was not popular. Nevertheless, it was revealed to be the case when Galileo Galilei aimed his primitive telescope at the sky in 1609.

In *An Original Theory or New Hypothesis of the Universe*, published in 1750, the surveyor Thomas Wright in Durham, England, argued that the Sun was either a star on a hollow sphere whose diameter was so enormous that we perceive only the local part of it as the Milky Way, or it was part of a hollow ring of stars;

he did not express a preference. Wright also speculated on theological grounds that there might be other such systems in the universe.

Immanuel Kant lived in Königsberg, Prussia, trained in theology, developed an interest in Newton's work, and later became an accomplished philosopher. In 1751 he happened upon a newspaper summary of Wright's book which gave the impression that the Milky Way was a disc-shaped system of stars. Kant found this idea appealing, and in 1755 he wrote the treatise *Universal Natural History and Theory of the Heavens*. The French

RIGHT Thomas Wright in 1737. *(Wikipedia)*

FAR RIGHT Thomas Wright's 1750 *An Original Theory or New Hypothesis of the Universe.*

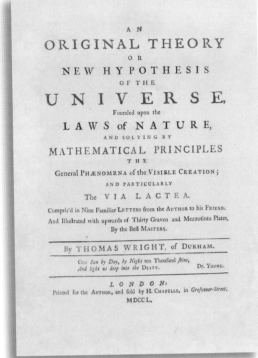

mathematician Pierre Louis de Maupertuis had written in 1742 of luminous patches that could be glimpsed by the naked eye, the most prominent being in Andromeda. Kant speculated that these were systems of stars of a similar character to the Milky Way and called them 'island universes'. Unfortunately, Kant's

publisher went bankrupt shortly before his book could be issued.

In 1764 the polymath Johann Heinrich Lambert told the Prussian Academy of Sciences in Berlin that the Milky Way was a disc-shaped system of stars, the Sun was situated towards its edge, and there were

RIGHT Immanuel Kant's 1755 *Universal Natural History and Theory of the Heavens.* *(W. D. Woods)*

FAR RIGHT Immanuel Kant.

providing an estimate of the relative dimensions and shape of the system of stars on the presumption that his telescope was capable of detecting stars right out to its edge. He inferred the Sun to be quite near to, but not precisely at the centre of the system.

Prior to Herschel, the celestial realm had been presumed to consist of the all-important Sun with its retinue of six planets, beyond which were stars, all at more or less the same distance. He had discovered a seventh planet and shown the Sun to be merely one star in a vast disc-shaped system of stars; not even occupying a special position within that system.

Most astronomers might have retired, satisfied with their achievements. Not Herschel. He continued to sweep the sky, adding to his catalogues. His long-term monitoring of what he called 'doubles' showed that in some cases the alignments were not simply by chance, as had been presumed – the stars were bound together and jointly revolving around a common centre. It was proof that Newton's law of gravitation applied universally.

By 1786, with a list of 1,000 nebulae, many of which were able to be resolved into stars,

Herschel suspected that with a sufficiently powerful telescope it would be possible to resolve all nebulae in this manner. He speculated that these groups of stars were so remote that they were independent systems, "some of which may well exceed our Milky Way in grandeur". He dramatically emphasised his point by suggesting that our system would appear as a fuzzy patch in the telescope of an astronomer on a planet orbiting a star in such a cluster.

Thus we see that years of painstaking observations, using superior telescopes that he created himself, enabled Herschel to relegate the immense star system that we perceive as the Milky Way to just one of many isolated star systems.

The next task for astronomers was to measure the distances to some stars in order to gain a sense of scale.

## A sense of scale

Cataloguing proper motions in 1792, Giuseppe Piazzi in Sicily found that star 61 Cygni was moving at 5.2 seconds of arc per annum, prompting its moniker of the 'flying star'.

**BELOW** The Dorpat Observatory (here shown in 1821) opened in 1810 with a state-of-the-art Fraunhofer telescope with a lens diameter of 9.5in.

Königsberg, Pr.     Sternwarte

Having fled Germany in the wake of Napoleon's invasion in 1808, Friedrich Georg Wilhelm Struve entered the University of Dorpat (now Tartu, Estonia). He became director of its observatory a decade later. The 9.5in refractor possessed the largest lens in the world, and proved to be ideal for making fine measurements of double stars. In 1830 Struve started to monitor 61 Cygni, a double separated by about 27 seconds of arc, and found that they revolved around a common centre in approximately 650 years.

When the Prussian government founded the Königsberg Observatory in 1810, Friedrich Wilhelm Bessel was appointed director. In 1818 he issued a catalogue of 50,000 stars, noting, where possible, their proper motions. In 1829 he installed a refractor whose performance was so good that he set about trying to measure the parallax of a star. He selected 61 Cygni on the presumption that, despite its being faint, its high proper motion required it to be nearby.

Ironically, determining this star's parallax was complicated by the fact that it was moving so rapidly, because it was first necessary to isolate the wobble in its track across the sky. Bessel asked his former assistant Friedrich Wilhelm August Argelander to search the archives and trace 61 Cygni's proper motion back to 1755. Within a few months of starting his observations in August 1837, Bessel was sure there was a displacemenet effect. To check for systematic effects, he finished the year and reported his success in December 1838. The angle of 0.31 seconds of arc placed the star at a distance of some 660,000au, the 'astronomical unit' being the radius of Earth's orbit around the Sun.

In 1831 Thomas Henderson, an amateur astronomer who earned his living as a legal clerk in Dundee, Scotland, was appointed director of the newly established observatory at the Cape of Good Hope in South Africa, the first major observatory in the far southern hemisphere. He began by updating the catalogue compiled by Halley from St Helena 150 years earlier. Upon finding that Alpha Centauri had a proper motion of 3.7 seconds of arc per year, Henderson realised that it must be nearby, an inference that was supported by the fact that it was the third brightest star in the sky. Furthermore, Alpha Centauri was a multiple system in which the motions of its components were readily discernible. This meant that they must be physically close to one another. Their separation in the sky was therefore another indication of the system's proximity. In April 1832 Henderson initiated a series of observations to try to measure the parallax of the main component.

In 1834 Henderson was made Astronomer Royal for Scotland, in Edinburgh. There, he analysed his Alpha Centauri observations and calculated a parallax of 0.9 seconds of arc. As a check he requested supplementary observations from the Cape, then published his result on 9 January 1839, several weeks too late to claim priority for being the first to achieve such a measurement, having been beaten by Bessel.

Given how enormously far away even the closest stars were, astronomers decided that the astronomical unit, an excellent measure of distances in the solar system, was too small for the stellar realm. The next obvious unit was the 'light year', this being the distance travelled by light in a year. As the speed of light is almost 300,000km/sec, the light year corresponds to a distance of about 10 trillion km.

In his study of the Milky Way, Herschel had estimated the diameter of its disc as 850 times the distance to Sirius. Once Sirius was calculated to be 9 light years distant, this calibration meant the diameter of the disc was a mind-boggling 7,500 light years.

By the mid-19th century, therefore, it was evident that the solar system, which a century earlier had appeared so large, was tiny in the context of the system of stars that we perceive as the Milky Way.

**BELOW The Observatory and Playfair's Monument on the summit of Calton Hill in Edinburgh, drawn by Thomas Hosmer Shepherd in 1883 and engraved by A. Cruse.**

Drawn by Tho. H. Shepherd.

Engraved by A. Cruse.

THE NEW OBSERVATORY, AND PLAYFAIR'S MONUMENT, CALTON HILL, EDINBURGH.

*Chapter Two*

# The Milky Way system

With the birth of astrophysics, astronomers began to accumulate masses of data on the motions and properties of stars. This provided an understanding of the formation and evolution of stars. The true enormity of the Milky Way system was determined by new methods for measuring stellar distances, and one particular kind of variable star which served as a 'standard candle'.

**OPPOSITE** A northerner's view of the Milky Way.
*(Ford Renton)*

RIGHT A portrait of
William Cranch Bond
in 1849 by Cephas
Giovanni Thompson.
(Harvard University)

BELOW The 'Great
Refractor' of the
Harvard College
Observatory had
a lens diameter of
15in. (Harvard College
Observatory)

# The birth of astrophysics

In the second half of the 19th century the advent of photography enabled the cataloguing of celestial objects on a scale far beyond what was possible by visual observations.

When Harvard College began to consider the construction of an observatory, William Cranch Bond, a prominent Boston clockmaker, was dispatched to Europe to inspect similar institutions there.

On his return, Bond submitted his report and created a private observatory to test techniques. When the Harvard College Observatory was opened in 1844, he was appointed as director. After a refractor with a 15in lens was installed several years later, the largest such instrument in the country and therefore known as the 'Great Refractor', Bond set about applying photography to astronomy.

Louis-Jacques-Mandé Daguerre, a French artist, had invented photography in 1830, but the very long exposures required for the 'Daguerrotype' process limited its use in astronomy.

Nevertheless, in 1840 the philosopher, physician and chemist John William Draper in New York was able to take a picture of the Moon using a simple lens to focus the image. A telescopic picture of the Moon taken by Bond in 1849 won a prize for technical excellence when it was displayed at the Great Exhibition in London several years later. The first photograph of a star, Vega, the brightest star in the constellation of Lyra, was obtained in 1850 by Bond working with prolific photographer John Adams Whipple.

The prospects for astrophotography were improved in 1851 when silversmith Frederick Scott Archer in England invented the 'wet' collodion process for glass-coated plates. The plate ceased to record an image when it dried out, so only short exposures were possible. Nevertheless, astronomer Warren de la Rue, inspired by Bond's image of the Moon, used the collodion process to obtain pictures of its face that were far superior.

Meanwhile, in the 1840s Friedrich Wilhelm August Argelander had led a team at the Bonn Observatory in Germany that visually charted the entire northern sky down to 9th magnitude. Between 1859 and 1862 they issued the Bonn Survey; the Bonner Durchmusterung, with 'BD' star designations. This astrometric catalogue with 457,848 stars was intended to provide quantitative data for future analysis.

In 1871 the English physician Richard Leach Maddox created 'dry' gelatine negative plates that could sustain much longer exposures to record faint stars. On becoming director of the Cape Observatory in South Africa in 1879, astronomer David Gill used gelatine plates to take pictures of the southern sky. The analysis of these plates took Jacobus Cornelius Kapteyn at the University of Gröningen in the Netherlands so long that the resulting Cape Photographic Durchmusterung did not become available until the final years of the century.

In 1835 the French philosopher Isidore Marie Auguste François Xavier Comte lamented that the constitution of the stars would "forever remain unknown". But in 1814 the Bavarian physicist Joseph Ritter von Fraunhofer, who was an expert in the properties of glass, had devised the spectroscope to study light in detail. Then in 1859 physicist Gustav Robert Kirchhoff and chemist Robert Wilhelm Bunsen of Heidelberg University opened a new 'window' into the celestial realm by using a spectroscope to identify elements in terms of spectral lines at certain wavelengths.

Hearing that spectroscopy could reveal the chemical composition of the stars, William Huggins in England compared this to "coming upon a spring of water in a dry and dirty land".

Born in 1824, Huggins developed a passion for astronomy as a teenager. He had planned to go to Cambridge but inherited the family drapery business, which he sold in 1854 in order to devote his time to the celestial realm. In 1858 he built an observatory by his house in London and installed a refractor with an 8in lens. On hearing of the work of Kirchhoff and Bunsen, Huggins put a spectroscope on his telescope. Observing visually, he wrote in his diary that almost every night's work was "red-lettered by some discovery". For a year, he investigated the light of the brightest stars. Although it was difficult to discern details, in 1863 he was able to report that the spectra of

**LEFT** Jacobus Kapteyn painted in 1921 by Jan Pieter Veth.

## NAKED-EYE STARS

Some 2,500 stars are visible to the naked eye, but that is the total for the entire celestial sphere and on any given night only a fraction of them are in view.

In the 2nd century BC, the Greek astronomer, geographer and mathematician Hipparchus of Nicaea (now İznik in Turkey) sailed to Alexandria in Egypt, which was a major scientific centre. After studying the sky, he compiled a catalogue of around 800 stars, noting their positions on a celestial grid that he conceived for the purpose, plus their brightness in terms of a system of 'magnitudes' numbered 1 to 6 in order of diminishing brightness (a rough and ready scale that allotted the 20 brightest stars to the first magnitude).

In 1856 Norman Robert Pogson in England formalised Hipparchus's scale by introducing a mathematical relationship in which a difference of five magnitudes corresponded to a 100-fold difference in brightness.

stars were 'continua' resembling the spectrum of the Sun.

While surveying the heavens, William Herschel had seen a number of circular nebulae. As each of these 'planetary nebulae' (as he described them) had a single faint star at its heart, he decided that the nebulosity must be a form of 'luminous fluid'. On examining a planetary nebula in 1864 Huggins found that its spectrum showed only a single glowing line of incandescent gas. He then looked at M31 in Andromeda and saw a continuum that meant the glow was produced by myriads of unresolved stars.

William Parsons was born in York, England, in 1800 as the son of Sir Lawrence Parsons. He was educated as Lord Oxmantown at Trinity College in Dublin, then graduated with a degree in mathematics from Magdalen College, Oxford, in 1822. As a financially independent 'gentleman astronomer', Parsons decided to create the most powerful telescope in the world in order to investigate the nebulae that had so fascinated William Herschel.

Dissatisfied by early long-focus refracting telescopes that suffered chromatic aberration, Herschel had perfected the art of making large mirrors for reflectors. Since then, astronomers had switched to the new achromatic refractors, but their lenses were of small diameter. To investigate the faint nebulae, Parsons required 'light grasp', so he opted for mirrors. In 1827 he began experimenting with casting processes, and in 1840 created a mirror with a diameter of 36in whose quality was so good that he set to work on a mirror with a diameter of 72in that would be his primary instrument. Following the death of his father in 1841, Parsons inherited the title of Third Earl of Rosse and took up residence at Birr Castle in Ireland.

In 1845 the 72in mirror was installed in a 56ft wooden tube that was slung on chains between a pair of stone walls in a transit-style mounting in the grounds of Birr Castle. The motion of this 'Leviathan' (as the telescope became known) was limited to 15° either side of the meridian, therefore an observer had to wait until a celestial object was near that line in order to view it.

Parsons' primary contribution to astronomy was the discovery that 15 nebulae are spiral

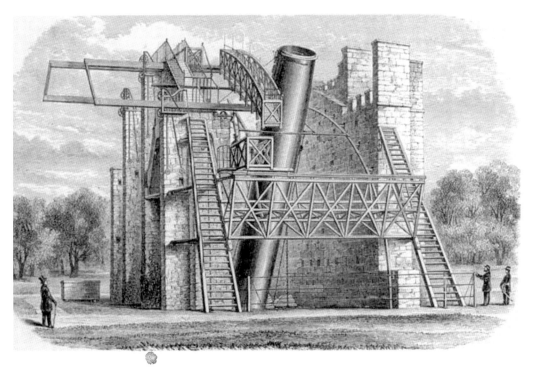

**LEFT** With a mirror
diameter of 72in,
the 'Leviathan' that
William Parsons built
at Birr Castle in Ireland
in 1845 was by far the
largest telescope in
the world at that time.
*(Wikipedia)*

in character. The most prominent was M51 in
the constellation of Canes Venatici, which he
named the 'Whirlpool' because the face-on
perspective makes its spiral most obvious.

The discovery that some nebulae are spirals
fully justified the effort involved in creating such
a massive telescope, and it caused quite a stir
in the astronomical community.

A century earlier, in 1755, Immanuel Kant
had suggested that the Sun and its retinue of
planets formed by a cloud of gas collapsing
under its own gravitational attraction. Pierre-
Simon Laplace was born in 1749, the son
of a farm labourer in Normandy. Noting his
intelligence, wealthy neighbours ensured that he
received an education. He went on to become

**LEFT** A sketch of
M51 by William
Parsons in 1845 as
he saw it through
his 72in reflector. Its
spiral form, which
he was the first to
perceive, led to it
being nicknamed the
'Whirlpool Nebula'.

RIGHT Pierre-Simon Laplace painted in 1775 by Johann Ernst Heinsius.

one of France's leading mathematicians. In 1796 he investigated how Kant's scheme might have occurred. He reasoned that because the Sun's rate of rotation would have increased as it contracted, it would have shed angular momentum by releasing a succession of co-planar rings which later condensed to form the planets.

Although Herschel (if he were still alive) would have intuitively reasoned that the spirals discovered by Parsons were systems of stars, they were assumed to be 'Laplacian nebulae' in which new stars were forming planetary systems.

By 1866, Huggins had examined 60 nebulae spectroscopically and found one-third to be

of the gaseous type; the remainder displayed continua typical of stars. The fact that the latter category included some of the spirals noted by Parsons meant they were not 'Laplacian nebulae' but systems of stars whose members could not be resolved. Herschel would have been delighted.

The Austrian physicist Christian Johann Doppler determined in 1842 that the wavelength of light should vary in proportion to the relative speed of its source, but it was 1848 before the details were refined by the French physicist Armand Hippolyte Louis Fizeau.

In 1868 Huggins visually estimated the displacement of a hydrogen line in the spectrum of Sirius, the brightest star in the sky, obtaining a velocity of 50km/sec. A survey of 30 other stars indicated that some were approaching and others were receding. However, it was difficult to make such measurements and these 'radial velocities' were only preliminary indicators.

Henry Draper, son of John William Draper, attended the School of Medicine of New York University. He finished the course at age 20 but had to wait another year to attain the minimum age for graduation. During that year, he indulged his interest in astronomy with a visit to Ireland to see the 'Leviathan'. On returning home he built his own observatory at his father's estate on the Hudson River. As a practical man, he created his own 15in mirror and the mechanism to accurately drive the telescope.

BELOW Henry Draper circa 1872.

BELOW RIGHT A diffraction spectrum of a star taken by Henry Draper.

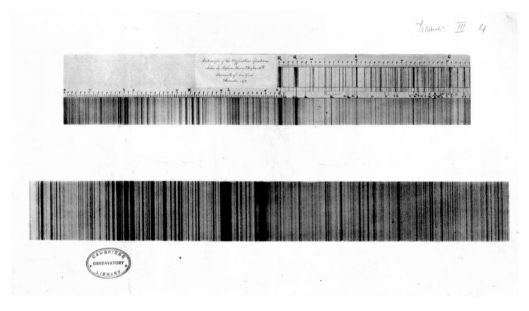

He started observing in 1861. In 1872, by now equipped with a 28in reflector, he succeeded in photographing the spectrum of Vega by the 'wet' collodion process despite the 15min constraint imposed by the drying out of such a plate.

When visiting Huggins in London in 1879, Draper was delighted to find that 'dry' plates had become available which were capable of much longer exposures and sufficiently sensitive for astronomical use. With these new plates on his large reflector he verified that some nebulae were glowing gas, and with exposures of up to 140min he captured stars that were too faint to be seen visually through the same telescope. In 1880 he obtained the first picture of the Orion Nebula. He also got excellent spectra of individual stars for a comparative analysis. Unfortunately, his efforts in this regard were cut short when he died of pneumonia in 1882 at the age of just 45.

In 1873 Hermann Carl Vogel at the Potsdam Astrophysical Observatory near Berlin, the world's first such facility, took up astrophotography. By 1888 he was routinely measuring the radial velocities of stars. In the early 1890s James Edward Keeler at the Allegheny Observatory in Pittsburgh began doing likewise.

By combining the proper motions of stars gained by photographic astrometry with their radial velocities determined from the displacement of the lines in their spectra, it was possible to calculate their real motions, their 'space motions'.

**LEFT** Edward Charles Pickering in 1912. *(Harvard College Observatory)*

Meanwhile, in 1885 Edward Charles Pickering, who was appointed director of the Harvard College Observatory in 1877, arranged for an 'objective prism' to be installed at the front of a telescope to produce a spectrum for each star in the field of view simultaneously, rather than photographing stellar spectra one by one. Of course the resolution was low, but it was enough to show general characteristics. He began a campaign to compile a comprehensive photographic star catalogue. In 1891 he sent his younger brother, William Henry Pickering, to operate a telescope at Arequipa, high in the Peruvian Andes, to extend the coverage into the southern sky.

**BELOW** The Arequipa Station in Peru established by the Harvard College Observatory, shown circa 1903. *(Harvard College Observatory)*

In 1867 Pietro Angelo Secchi at the Observatory of the Roman College in Italy had devised the first classification scheme for stellar spectra on the basis of his own studies. Meanwhile, in Sweden the physicist Anders Jonas Ångström had spent a decade studying the Sun and in 1868 he published a chart of 800 lines in the solar spectrum. After comparing the stellar spectra obtained by Draper to those of the Sun, E. C. Pickering introduced a classification that subdivided the one offered by Secchi.

Dissatisfied with the performance of male assistants in analysing the objective prism spectra, Pickering assigned the task to his Scottish housemaid, Williamina Fleming, because he reckoned a woman would be more fastidious. He soon began hiring female assistants, known as 'computers', to assist with the growing task of classifying stellar spectra.

Antonia Cætana Maury was the granddaughter of John William Draper and a niece of Henry Draper. She was hired by Pickering immediately after graduating from Vassar College in 1887 in physics, astronomy and philosophy.

Annie Jump Cannon graduated from Wellesley College in 1884 with interests in physics and astronomy. To further her education and gain access to telescopes she enrolled at the Society for the Collegiate Instruction of Women. Founded in 1879 as an annexe to Harvard, in 1894 it became Radcliffe College. Its *raison d'être* was to enable professors to repeat their lectures to women, as Harvard was male-only. Cannon joined Pickering's team in 1896.

Most of the photography and analysis was funded by Draper's widow, under the stipulation that the result be called the Henry Draper Catalogue. Starting in 1890, various lists were published that progressively expanded the coverage. The final product in the early 20th century gave spectroscopic classifications for over a quarter of a million stars. It encompassed the entire sky down to a photographic magnitude of about 9. This monumental effort provided the basis for the science of astrophysics.

# Variable stars

**B**orn in 384 BC, Aristotle was a student of Plato at the Academy in Athens, at that time the leading centre for philosophy. He went on to write books on a wide range of topics that profoundly influenced later scholars. A fundamental tenet of Aristotle's philosophy was the 'perfection' of the celestial realm. It was therefore axiomatic that the stars remained in position and maintained a given brightness. The comets that sometimes appeared in the sky were deemed by Aristotle to be in the upper atmosphere and therefore not part of the celestial realm that he held in such esteem.

When the classical astronomer Hipparchus saw a 'new' star suddenly appear in 134 BC, slowly fade, and finally vanish, he compiled a catalogue of 800 stars to assist his successors in recognising such an event in the future.

Yet so sure were Europeans that the celestial realm was unchanging that they paid no attention to such apparitions. They made no record of the appearance in July 1054 of what Chinese chroniclers described as a 'guest star' that was initially visible in daylight for several weeks and subsequently at night in the constellation of Taurus.

Fortunately, in the middle of the second millennium people in Europe became more observant. When a 'new star' appeared in Cassiopeia on 11 November 1572, outshone Venus for several weeks, then faded and disappeared in early 1574, the Danish nobleman Tycho Brahe published his observations in a pamphlet entitled *De Nova Stella* (*The New Star*). This finally awakened Europeans to the fact that the celestial realm was not without change.

## MIRA

When clergyman David Fabricius in Germany saw a star in Cetus brighten by 1 magnitude over the period of several weeks in August 1596 and fade from view by October, he presumed it to be similar to previous stars that appeared and then vanished. However, it was seen again in 1603 by Johann Bayer, a German lawyer, who listed it as Omicron Ceti in an atlas of the sky that he was compiling. Then it disappeared again.

After several years of observations, in 1638 the Frisian astronomer, physician, and philosopher Johannes Phocylides Holwarda realised that this particular star was varying across a range of about 9 magnitudes in a period of about 330 days. The 'light curve' was a pseudo-sinusoidal wave that rose somewhat faster that it decayed and followed a general pattern without ever precisely repeating. In view of the presumed uniqueness of this 'variable star', in 1662 the German astronomer Johannes Hevelius named it 'Mira' (meaning 'the Wonderful').

## ALGOL

The first recorded statement of the variability of the star Algol, listed as Beta Persei, was made by the Italian lawyer and astronomer Geminiano Montanari in 1667, but the fact that the star was widely known as the 'Demon Star', referring to a gorgon or ghoul, implies its variability was recognised much earlier.

The periodic nature of Algol's variability was discovered by John Goodricke, a deaf-mute teenager living in York, England, who had a passion for astronomy. In 1783 he reported to the Royal Society his inference from the manner in which the light curve repeated (dipping from magnitude 2.1 to 3.4 every 2.86 days for about 10hr) that either the star had a dark patch which was sometimes turned toward us as the star rotated axially, or the star was accompanied by a faint companion that sometimes obscured our

BELOW A painting by Raffaello Sanzio in 1509 of 'The School of Athens' has as its main characters Plato (left) and Aristotle, the greatest thinkers of classical antiquity. The original (of which this is merely one part) is currently in the Vatican.

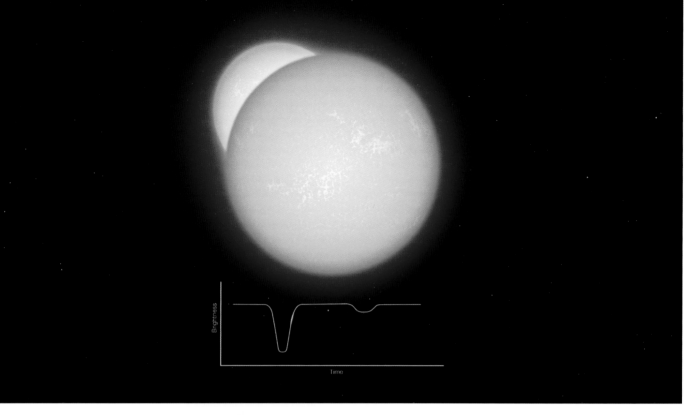

ABOVE An explanation of an eclipsing binary star system. *(ESO/L. Calçada)*

RIGHT John Goodricke painted in 1785 by James Scouler.

RIGHT The light curve of the star Delta Cephei. Its brightness oscillates in a regular manner between well-defined maxima and minima over a period of 5.4 days. *(W. D. Woods)*

view of the bright star. The binary nature was confirmed a century later, making it the first example of a class of variables called 'eclipsing binaries'.

## DELTA CEPHEI

In 1784, shortly before his death at the age of only 21, Goodricke discovered that Delta Cephei has a period of 5.4 days. Its brightness rapidly rises to a given peak magnitude and then slowly fades to a given minimum. Over the years, other 'Cepheids' (as they became known) were found with periods ranging from about a day to several months; the mean being about a week.

Solon Irving Bailey joined the Harvard College Observatory staff in 1886 as an unpaid assistant, but his work was of such a high calibre that he was soon given a salaried

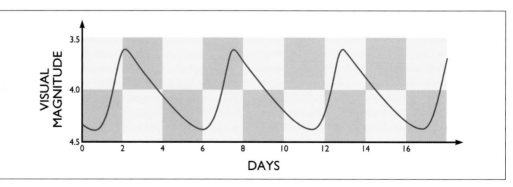

position. After being assigned to Harvard's station at Arequipa in Peru in 1893, he started to photograph globular clusters, extremely dense clusters of stars, of which there are many in the southern sky. He placed a transparent grid over an image of the cluster named Omega Centauri and counted the stars in each square to measure their distribution. The dense central region could not be resolved, but he was able to count over 6,000 stars surrounding it. With so many stars in such close proximity, a globular offered rich pickings for variables. In addition to some Cepheids with periods of 12 to 20 days, there were many similar but fainter RR Lyrae variables that cycled in about 12hr. He coined the term 'cluster variables' for this short period variety.

It was not immediately evident that such stars would play a significant role in discovering the true scale of the universe.

**LEFT** Solon Irving **Bailey.** *(National Academy of Sciences)*

**LEFT** The globular cluster Omega Centauri viewed from the European Southern Observatory at La Silla in Chile. *(ESO)*

# A 'standard candle'

**H**aving attended a general course on astronomy at the University of Missouri in 1907, Harlow Shapley gave up his plan to train as a journalist and pursued this new interest instead. He received his doctorate from Princeton University in 1914 for a study of eclipsing binaries under the supervision of Henry Norris Russell, a leading figure in the study of stellar spectra.

As Shapley was planning his post-doctoral work, George Ellery Hale, director of the Mount Wilson Observatory in California, invited him to make use of the 60in reflector, at that time the most powerful functioning telescope in the world. Shapley accepted. On a visit to Harvard in March 1914, Shapley met Bailey, who upon learning that the visitor was joining Hale's staff, said: "When you get there, why don't you use the big telescope to measure stars in globular clusters?"

On 13 November 1833 John Herschel sailed from Portsmouth for the Cape of Good Hope, arriving several months later. Once established, he set up his father's favourite 20ft reflector and began a detailed survey of the southern sky. However, he was easily distracted, most notably by the local botany, and the progress of his astronomical work was slow.

When the explorer Ferdinand Magellan sailed into the far southern ocean in 1519 he noted in his log the presence of two bright patches in the sky that looked just as if they were fragments of the Milky Way that had drifted loose. Although his vessel, *Victory*, completed the historic first circumnavigation of the globe a few years later, the man himself was killed in battle in the Philippines and the celestial clouds were named in his honour. Herschel paid them particular attention, and was able to resolve not only individual stars but also clusters of stars and patches of glowing nebulosity.

Back in London in 1838, Herschel set about processing his observations, and in 1847 issued a catalogue that increased to almost 100 the list of globular clusters. Their concentration near the plane of the Milky Way implied they were related to that system, yet, astonishingly, about 40% of them occupied less than 3% of the sky in the direction of Sagittarius.

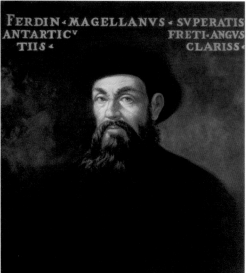

**FAR LEFT** This portrait of John Frederick William Herschel is an engraving by William Ward of a painting by H. W. Pickersgill.

**LEFT** The early 16th century explorer Ferdinand Magellan. The portrait is in the Uffizi Gallery in Florence, Italy.

In setting out to investigate this asymmetric distribution, Shapley focused his attention on the RR Lyrae variables. Their cycles were so rapid that when Bailey had taken long exposures on a small telescope to follow them as they faded down to their minima, the loss of temporal resolution had made it difficult to precisely trace their light curves. The much greater light grasp of the 60in telescope enabled Shapley to follow them in detail.

Realising that Cepheids were not eclipsing binary stars, Shapley investigated how a star might pulsate. He lacked a detailed theory for why this might happen, but managed to show mathematically that the period of pulsation would depend only on the density of the star. Specifically, the product of the period of pulsation and the square root of the density would be the same in all cases. This meant that a star which was less dense would pulsate with a longer period.

When Cepheid variables were monitored spectroscopically, it was found that the spectral lines were shifted toward the blue end of the spectrum when the stars were at their brightest and toward the red end at their faintest. These oscillations confirmed the stars were pulsating, with their radius inflating and contracting by as much as 10% in only a few hours.

The distances to around 60 stars had been determined by 1900, essentially by the same method as that used by Bessel, but this was possible only for stars which were relatively nearby. The accuracy of direct parallax

**BELOW** Globular clusters (circled) in the constellation of Sagittarius. One third of all known globulars are in this frame, which covers only about 2% of the sky. *(Harvard College Observatory)*

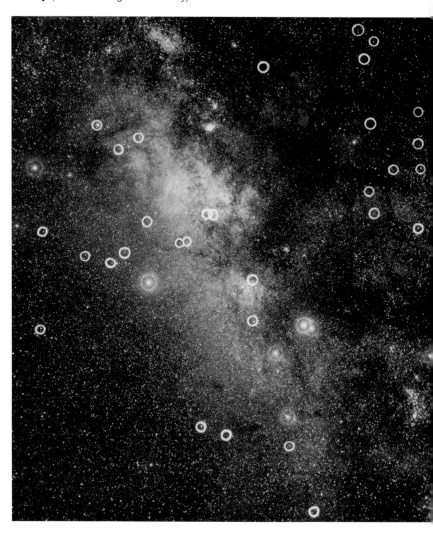

## STATISTICAL PARALLAX

With radial velocities determined from spectra, astronomers had, for the first time, a quantity that was not influenced by distance, so they combined this with a quantity that did depend on distance as a way of determining distances.

In the half century after Halley had reported that some stars were moving on the sky, only 14 proper motions had been accurately measured. Nevertheless, in 1783 William Herschel presented his interpretation in a paper to the Royal Society 'On the Proper Motion of the Sun and Solar System'. He found that the Sun was travelling through space toward the 'solar apex' in the constellation of Hercules. By 1900, new data confirmed Herschel's analysis to be broadly correct.

Of course, the Sun is not unique in moving through space. All stars have their own individual 'space motions'. By presuming their velocities to be random (not necessarily the case, but a fair assumption in the absence of data to the contrary) it was possible to estimate distances by taking the ratio of random speeds through space to the random apparent motions on the sky. That is, for a group of stars that were of the same spectral class and a similar magnitude, implying they were at a similar distance, it was possible to derive a mean parallax by a statistical analysis of their proper motions relative to their radial velocities. As catalogues of data for stars became available, they facilitated 'number crunching' studies of this nature.

This form of statistical parallax is useful for estimating the distances of bright stars that are beyond direct triangulation range.

measurements was greatly improved by the introduction of astrophotography, but by the turn of the century it was evident that indirect methods would have to be devised in order to achieve a deeper understanding of the situation.

Fortunately, with proper motions and radial velocities becoming available for a large number of stars, it was possible to identify the actual motions of stars in space. As William Herschel had noticed, there was a general drift in which stars in widely different regions of the sky shared a common motion. Of course, not all stars followed precisely the same vector. The degree to which the true velocity of a star differed from the mean could be translated into a unique motion across the sky and interpreted as a measure of parallax. Since this and other such methods were of statistical form, the result was called 'statistical parallax'.

Another method involved star clusters. Richard Anthony Proctor had noted in 1869 that the proper motions of five of the stars

**BELOW** The proper motions of the Plough in the constellation Ursa Major show that most of the stars share a common motion.

*(Image A. Fujii/annotation W. D. Woods)*

*Chapter Three*

# Discovering creation

In exploring the cosmological significance of his theory of General Relativity, Albert Einstein accepted the advice of contemporary astronomers that the universe had always existed and was unchanging. But further studies have established that not only are 'spiral nebulae' systems of stars comparable to our own Milky Way system, but also that their systematic motions mean the universe is expanding.

**OPPOSITE A wide-field view of the M31 spiral in Andromeda by a terrestrial telescope for the Digitised Sky Survey in support of NASA's Space Telescope Science Institute.**
*(Adapted from NASA/ESA/DSS2/Davide De Martin)*

RIGHT Isaac Roberts.

RIGHT Isaac Roberts.

# The issue of the spirals

In 1885 Carl Ernst Albrecht Hartwig, a German astronomer at the Observatory of the Imperial University of Dorpat in Estonia, saw a 'new star' which, although it peaked at an unremarkable 7th magnitude, was significant because it was positioned in the middle of the M31 nebula in Andromeda. In 1796 Pierre-Simon Laplace in France had proposed a hypothesis for how planetary systems formed as stars condensed out of clouds of gas. It was now argued that M31 must be a 'Laplacian nebula' in which the central star had just 'switched on'. Its subsequent disconcerting fading from sight was explained as a swirl of dust masking the star.

The English mechanical engineer and amateur astronomer Isaac Roberts was an early enthusiast of 'dry' plates for astrophotography. After experimenting in 1883 with portrait lenses, he built an observatory and bought a telescope with an equatorial mount. He then affixed his astrograph alongside and viewed through the telescope to maintain the aim during exposures lasting an hour or more to record faint objects. In 1886 he presented his first photographs at a meeting of the Royal Astronomical Society, revealing "the vast extensions of nebulosity in the Pleiades and Orion".

In December 1888 Roberts achieved an excellent picture of M31. To date, this fairly bright nebula had been described as a fuzzy oval, but Roberts revealed it to be much larger, with a prominent spiral shape in its outer region. This structure may have eluded the visual observations of William Parsons because the field of view of his 'Leviathan' was so narrow. In line with Laplace's rationale, Roberts believed the nebula must be "a new solar system in the process of condensation".

But then in 1899 Julius Scheiner in Germany took a spectrum of M31, with an exposure of almost 8hr, whose continuum implied it

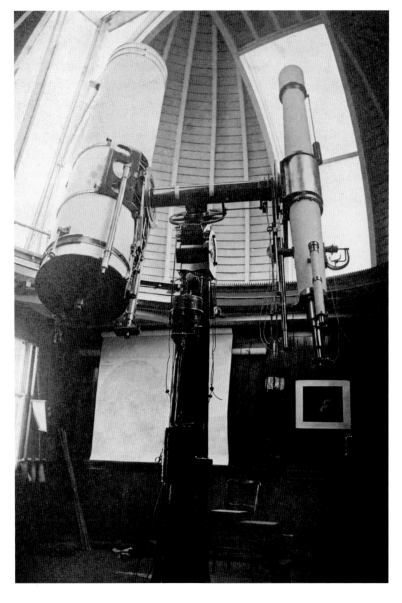

LEFT The apparatus with which Isaac Roberts obtained his photograph of M31. During the long exposure, he would use one telescope to track the target while the plate was exposed using the other instrument.

comprised unresolved stars. Was it a small cluster within the Milky Way system? Or an independent system comparable in size to our own? If the latter, then it must be incredibly far away.

At the Lick Observatory, James Edward Keeler began to search for nebulae in 1898 by photographing selected portions of the sky. In the half-century since M51 was revealed by Parsons to have a spiral form, only a few dozen other spirals had been listed. Keeler discovered so many spirals that he concluded there must be at least 100,000. The followers of Laplace took this number to mean that the spirals must be part of the Milky Way system, since it was "absurd" to imagine that there could be so many independent comparable systems.

On graduating from Harvard, Percival Lowell toured the world and was then assigned to run one of the family businesses. However, he had developed such a passion for astronomy that in 1893 he decided to build his own observatory near the small sleepy railroad town of Flagstaff in the Arizona Territory, which, being on a plateau at 7,200ft, offered excellent 'seeing' conditions for much of the year. He started observing in April 1894 using a borrowed 18in refractor, but replaced

**BELOW** The historic photograph of M31 taken by Isaac Roberts in December 1888, revealing not only the true extent of the nebula but also that its outer region had a spiral form.

ABOVE The dome of the 24in refractor erected at Flagstaff in the Arizona Territory by Percival Lowell in 1894. *(Lowell Observatory)*

result of reflecting the light of the Pleiades stars which it contained, critics of the idea that spirals were island universes argued that M31, which also had a continuum spectrum, must be similar and so dense that its gas entirely masked the stars within.

Believing the spirals to be Laplacian nebulae, Lowell gave Slipher the task of measuring the radial velocities of such nebulae in order to determine their rate of rotation. This posed a considerable challenge. Lengthy exposures were necessary simply to photograph spirals, and it would take even longer to record their feeble light when it was dispersed to form a spectrum. Measuring the radial velocity of a spiral by the Doppler shift required being able to discern lines in the continuum, but as yet no one had been able to resolve spectral lines.

In 1909 Slipher began photographing the spectrum of M31, because it was by far the brightest spiral in the sky. The results were disappointing. The 'speed' of the camera installed behind the prism on the 24in refractor was too 'slow', so he adapted a faster, short-focus camera. By September 1912 he was making progress, and by January 1913 he had four plates showing measurable lines whose Doppler shifts toward the blue end of the spectrum indicated a radial velocity toward us of about 300km/sec.

this in 1896 with a new telescope that had a 24in lens.

In 1901 Lowell hired Vesto Melvin Slipher, a recent graduate of the University of Indiana, to investigate the spectra of nebulae. When Slipher found M45 to have a continuum spectrum as a

As this velocity was so much greater than had been measured for any other celestial object, it attracted scepticism from William Wallace Campbell at the Lick Observatory, who was regarded as the authority on radial velocities. But Slipher persevered, and in the paper 'Spectrographic Observations of Nebulae' presented to the American Astronomical Society in August 1914, he gave radial velocities for 15 spirals. Two were approaching but the others

RIGHT Vesto Slipher with the spectrograph on the 24in refractor of the Lowell Observatory. *(Lowell Observatory)*

### REDSHIFT

The Doppler effect displaces spectral lines by an amount proportional to the velocity along the line of sight, called the radial velocity. In the case of light from an object that is moving away from us, the lines are displaced toward the red end of the spectrum; hence the term 'redshift'.

were receding, two of them at over 1,000km/sec. His presentation earned him a standing ovation. A few months later, Campbell accepted this was "one of the greatest surprises that astronomers have encountered in recent times". Slipher's astronomy mentor at Indiana University, John A. Miller, said the discovery opened up a "gold mine".

By 1925 Slipher had 41 measurements, and astronomers elsewhere, having made their own 'fast cameras', had obtained 'redshifts' for four other spirals. The greatest redshift corresponded to the astounding velocity of 1,800km/sec. But by then, Slipher was at the operating limit of the 24in refractor and could proceed no further. Following Lowell's death in 1926, Slipher was appointed director of the renamed Lowell Observatory and he turned his attention to other projects.

## A theory of the universe

**A**fter studying astronomy under Jacobus Cornelius Kapteyn at the University of Gröningen in the Netherlands, Willem de Sitter obtained his doctorate from the University of Leiden in 1901 where, in 1908, he was appointed both professor of theoretical astronomy and director of the Leiden Observatory.

Upon graduating from the University of Cambridge in 1907, Arthur Stanley Eddington was hired by the Royal Greenwich Observatory.

**ABOVE LEFT William Wallace Campbell in 1901.** *(Lick Observatory)*

**ABOVE Willem de Sitter.** *(University of Chicago)*

Five years later, he returned to Cambridge as professor of astronomy and experimental philosophy, and soon became the 'go to' person for research insights.

Eddington and de Sitter corresponded

**LEFT Arthur Stanley Eddington.** *(Smithsonian Institution)*

regularly on issues arising from Albert Einstein's Special Theory of Relativity, published in 1905, and although the Great War was raging when Einstein's paper 'The Foundations of the General Theory of Relativity' was published in Germany in 1916 stating the relationship between the contents of space and the geometry of space, the Netherlands was neutral and de Sitter was able to mail a copy to Eddington.

In developing General Relativity, Einstein rejected Newton's explanation of gravity as action at a distance, seeing it instead as the curvature of 'space-time' by the presence of mass and energy. To model it, he employed the mathematical concept of a 'field', in which a property has a value at every point in space.

The 'field equations' did not set boundary conditions, but Einstein took it for granted there would be a single solution and, having discovered it, in February 1917 he published his analysis in the paper 'Cosmological Considerations of the General Theory of Relativity'. He assumed the universe is isotropic (meaning it looks the same in every direction) and that it looks the same from every vantage point (meaning we don't occupy a privileged position).

On discovering that the equations indicated the universe could not be static, in the sense that, overall, things wouldn't change with time, instead requiring it to be either expanding or contracting, Einstein asked astronomers and was assured that the Sun was situated near the centre of a disc of stars that was self-evidently stable. Einstein added a new term into his equations, the 'cosmological constant', to prevent the universe from collapsing under its own gravitational attraction. It imbued the vacuum of space with an intrinsic pressure that manifested itself as a repulsive force throughout the universe, with a magnitude calculated to precisely counter self-gravitation. In effect, he was invoking antigravity. However, the only 'evidence' of this force was the presumption that it must be holding the universe stable against collapse.

After reading Einstein's 1916 paper about General Relativity, de Sitter carried out a correspondence with Einstein and then wrote three papers of his own. In the third, in 1917, he argued that the stars are so far apart that the universe is mostly empty space. Because the 'matter density' is insignificant, for the purpose of his analysis he presumed it to be zero. He retained the cosmological constant because he sought a static solution. Although his universe was devoid of matter, when de Sitter added two mathematical 'test particles' he found that light passing between them would be redshifted and that the redshift would be directly proportional to their separation. However, because de Sitter believed his universe to be static, he interpreted this redshift as the slowing down of time across great distances; this became known as the 'de Sitter effect'.

But then, when Eddington examined de Sitter's solution, he realised that the space which the equations described was in a state of expansion and the redshift was a manifestation of divergence. In response, de Sitter quipped that his was a static solution in the sense that there was no matter to reveal its expansion!

Einstein rejected de Sitter's solution as lacking physical significance. After all, the universe was not empty.

Eddington was intrigued that Einstein could identify a solution to the field equations in which there was matter but no expansion, while de Sitter found a universe with no matter that was nevertheless in a state of expansion.

The surprising thing was that it seemed that no other solutions were possible under the assumption that the universe was static. As a result, the two solutions by Einstein and de Sitter were widely studied in an effort to determine which one corresponded to reality. But events in the newly established Soviet Union were to upset the applecart…

In 1910, Alexander Alexandrovich Friedmann joined the mathematics faculty at the University of St Petersburg. After the Great War and the Revolution, he accepted a teaching position at the university in Perm. In 1920 he returned to St Petersburg, which by then was Leningrad. Although a meteorologist, Friedmann had an interest in Einstein's work. In 1922 Friedmann ignored the focus on static solutions and studied a family of expanding solutions, one being 'open', one 'flat', and the third 'closed', distinguishable in terms of the matter density. His results were published in two papers in 1922 and 1924 in a recently established German scientific journal.

Rather than calculate, as Einstein did, a particular value of the cosmological constant to uphold the assertion of astronomers that the universe was static over time, Friedmann had derived solutions for a range of values of the cosmological constant. He had one in which the cosmological constant was zero,

which was essentially what Einstein had prior to forcing it to be static. Friedmann allowed the possibility of a dynamic universe that was either expanding or collapsing. He identified three possible outcomes, depending on the initial rate expansion and the mean matter density. If there was sufficient mass, then self-gravitation would eventually halt the expansion and initiate collapse. If the density was insufficient to halt expansion, the universe would expand for ever. At the critical density, the expansion would be slowed but would require an infinite time to come to a halt. Thus Friedmann showed that if the universe was in a state of expansion and the rate of expansion was sufficiently great, then it need never collapse – even if the cosmological constant was zero. Unfortunately, Friedmann was unable to pursue his insights because he died in 1925 after contracting an infection.

On learning of Friedmann's results, Einstein was surprised that his equations could generate so many solutions. However, because astronomers still favoured a static universe he rejected the possibility that it might be in a state of expansion.

During the war, Georges Lemaître served as an artillery officer in the Belgian army. Then he attended the University of Louvain and was ordained as a priest in 1923. In working for his doctorate he spent a year with Eddington

in Cambridge and a year in Boston, attending both Harvard, where he was under the tutelage of Shapley, and the nearby Massachusetts Institute of Technology. On returning to Belgium in 1925, Lemaître worked as a lecturer at his alma mater while writing his thesis. In 1927 he was appointed professor of astrophysics.

In exploring Einstein's field equations, Lemaître was unaware of Friedmann's work (as indeed was Eddington). He independently discovered that the universe could be in a state of expansion but, unlike his predecessor, Lemaître had a better knowledge of astronomy. While in America he heard of Slipher's radial velocities for spiral nebulae, so he was more readily able to accept that the universe might be expanding. He realised that light from

distant objects would be redshifted as a result of the expansion of the intervening space, and predicted that the degree of redshift would be directly proportional to the distance. Lemaître initially avoided the issue of an 'origin' by retaining the cosmological constant and presuming that, after spending a long time in the static state envisaged by Einstein, an instability of some sort had triggered the runaway expansion that we see today.

Lemaître's paper, 'A Homogeneous Universe of Constant Mass and Growing Radius Accounting for the Radial Velocity of Extragalactic Nebulae', appeared in the *Annals of the Brussels Scientific Society* in 1927. He sent a copy to Eddington but the latter seemingly mislaid it.

At the Fifth Solvay Conference in Brussels in October 1927, where the leading physicists discussed the newly emerging Quantum Theory, Lemaître encountered Einstein and gave him a précis of his work. Einstein said he had seen the papers by Friedmann with solutions to the field equations which allowed the universe to expand (thereby introducing Lemaître to this earlier work) and curtly dismissed Lemaître, saying, "Your calculations are correct, but your physics is abominable."

Lemaître sent Eddington a second copy of his paper after learning that at a meeting of the Royal Society of London in January 1930, the Englishman had raised the question of how

Einstein's solution to the field equations might require to be revised in view of the redshift-distance relationship recently announced by Edwin Hubble (see below).

As it happened, Eddington had just set a student, George McVittie, the task of studying General Relativity in terms of an expanding universe, so was delighted to receive Lemaître's paper explaining it. Eddington published a note in *Nature* in June 1930 drawing attention to Lemaître's work and describing it as "brilliant". He also arranged for the paper to be translated from French into English and had it published by the Royal Astronomical Society.

Rejecting the idea that the universe was initially in a static state and started to expand only in response to an instability, Lemaître realised that if one could 'wind the clock back' there must have been a 'time zero' when all matter in the universe emerged from a single superdense entity. He called this the 'primeval atom' and suggested that it had disintegrated by a process analogous to that of spontaneous radioactive decay. What we see today, he said, is the debris scattering. This meant that General Relativity predicted a moment of creation. Being a priest as well as an astronomer, Lemaître found this idea appealing. He published this research in 1931 in 'The Beginning of the World from the Point of View of Quantum Theory', this time in *Nature* in order to reach the widest possible audience.

In retrospect, it is interesting that whereas Lemaître imagined the expansion of the universe starting with the disintegration of the 'primeval atom', an entity which was difficult to envisage, Friedmann had conceived the universe emerging from a dimensionless point, a mathematical singularity, that was even harder to imagine. However, as we shall see, the impossible-sounding singularity proved to be closer to the truth!

## A staggering sense of scale

In July 1917 George Willis Ritchey was comparing two plates taken using the 60in telescope at Mount Wilson of a spiral in Cepheus when he noticed that a star in its outer fringe was present on one plate and not on the other. In this peripheral location, it could not be a newly formed star in a Laplacian nebula. Monitoring of the distinctive light curve established it to be a nova. Ritchey then examined the observatory's archive and found two novae in M31 on plates that he had taken in 1909.

By 1919, systematic surveys had found another 14 novae in M31. They were all similar to one another and utterly different from that seen in the same nebula in 1885. This implied that the latter could not be compared with local novae and therefore all distance estimates based on it were invalid. After analysing the new sample of novae in M31, Heber Doust Curtis of the Lick Observatory estimated it to be about 500,000 light years away, which made it an independent system with millions of stars.

The addition of the 100in telescope at Mount Wilson improved the prospects for studying the nature of the spirals, but even with its unrivalled power it was difficult to be sure that individual stars were being resolved. When a point which equated to half of a second of arc on a photographic plate was scaled to Curtis's estimated range for M31 of half a million light years, it spanned over a light year, a volume of space sufficient to accommodate a dense cluster of stars.

For its 1920 meeting in Washington DC, the National Academy of Sciences decided to feature two speakers who would offer differing opinions on 'The Scale of the Universe'. They were Harlow Shapley at Mount Wilson, who had recently published a study of the distribution of globular clusters which suggested that the Milky Way system was so large that it must surely constitute the entire universe, and Curtis at Lick, the leading advocate for the 'island universe' theory in which the spiral nebulae were comparable to the Milky Way system. By chance, the two men found themselves crossing the nation on the same train, and by an unspoken agreement they steered clear of astronomy in their conversations.

The event was hosted by the Smithsonian Institute on the evening of 26 April. After the Academy had given awards to scientists who had recently made major contributions to their fields, there was a dinner. The presentations were to draw the evening to a close. The audience consisted of about 200 people who, owing to the number of disciplines represented, were by no means astronomically literate. After each 40min talk there was time for audience questions but no direct rebuttal by the protagonists, so it wasn't a debate.

Shapley spoke first, reading a script intended for an intelligent layperson in which he spent several minutes introducing the term 'light year'. Picking up the pace, he explained the size and general composition of the Milky Way system, but deliberately omitted the technical details of the processes by which measurements were achieved. He concluded that if spirals were systems of stars they couldn't be systems comparable with our own.

Then Curtis presented a technical argument, saying that Shapley had greatly overestimated the dimensions of the Milky Way system. He cited evidence that it had a spiral structure. The radial velocities of spiral nebulae were far greater than any objects within the Milky Way system and couldn't be part of it. Therefore the universe consisted of a large number of comparable systems distributed across an enormous void.

It had been agreed that the speakers would each submit a formal paper to be published together in the *Bulletin* of the National Research Council. Shapley and Curtis corresponded while writing their papers. Curtis provided an updated form of his talk, but Shapley's paper bore almost no relation to his oral presentation. In contrast to the talks, the published papers represented a structured debate. People who hadn't attended got the impression from the papers that it must have been a real fireworks display! Hence the retrospective title of the 'Great Debate'.

The essence of the dispute was whether Shapley's distances for the globulars were plausible, so the Debate in the papers published in 1921 wasn't really about the nature of the spirals, it was about the size of the Milky Way system. Although Shapley and Curtis disagreed vigorously in public, neither man's career suffered. Indeed, shortly afterward Curtis became director of the University of Pittsburgh's Allegheny Observatory in Pennsylvania, and

# GEORGE ELLERY HALE

George Ellery Hale was born in 1868. His father made a considerable fortune manufacturing and installing passenger elevators in the reconstruction of Chicago following the Great Fire of 1871. He graduated from the Massachusetts Institute of Technology in 1890 with a degree in physics, in his final year developing the spectroheliograph to observe the Sun.

Hale talked the Chicago businessman Charles Tyson Yerkes, who had made a fortune installing mass-transit, into financing the construction of a telescope with a 40in lens, the largest in the world, and an observatory to accommodate it. When the Yerkes Observatory was created in 1897 near the town of Williams Bay, north of Chicago, Hale was appointed its first director.

In 1904 Hale took a small telescope to the 5,700ft summit of Mount Wilson in the San Gabriel Mountains east of Los Angeles to assess the 'astronomical seeing'. When Hale resigned from Yerkes in 1905, he was succeeded as director by Edwin Brant Frost.

The solar telescope that Hale erected on Mount Wilson with funding from the Carnegie Institution was only the start. He was already planning a telescope with a 60in mirror purchased by his father, with the observatory to house it funded by Carnegie. The 'first light' for this was in December 1908 and it was commissioned in 1909. Although smaller than the 72in built by William Parsons in Ireland, the 60in was more advanced, not least because its mount enabled it to view any part of the sky rather than the narrow zone near the local meridian. And of course, the 'Leviathan' had long since been retired.

In 1906 Hale convinced the Los Angeles businessman, amateur astronomer, and philanthropist John Daggett Hooker to fund a 100in mirror and Carnegie to pay for the telescope. It was ready for testing in late 1917 but with the observatory staff directing their attentions to other work during America's involvement in the Great War, the telescope didn't become available for general use until September 1919.

**LEFT** George Ellery Hale circa 1912.
*(Wikipedia)*

**BELOW** With a lens diameter of 40in, the Yerkes refractor is still the largest in the world.
*(Yerkes Observatory)*

ABOVE **The dome of the 100in telescope on Mount Wilson.** *(Observatories of the Carnegie Institution)*

Early on, Hale had hired Walter Sydney Adams, who had studied astronomy under Frost in Chicago, to assist with the establishment of the facilities at Mount Wilson.

By 1916, Hale was spending so much time on other business that in 1917 he made Adams his deputy to handle the day to day administration. When Hale resigned in 1923

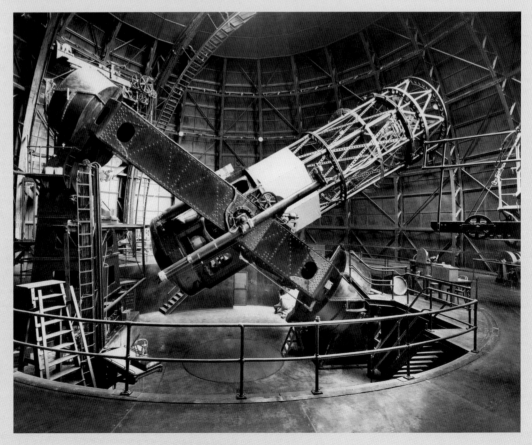

RIGHT **With a mirror diameter of 100in, the Hooker telescope on Mount Wilson was the largest in the world.** *(Observatories of the Carnegie Institution)*

due to persistently poor health, Adams took over as director and served in that role through to 1946.

For his finale, Hale got the Rockefeller Foundation to fund the construction of the 200in reflector on Mount Palomar, south-east of San Diego in California. It was intended to complete it in the early 1940s, but progress was slow and then paused during the Second World War. Finally, in 1948, over 1,000 people gathered in the vast dome for the dedication of the new telescope to George Ellery Hale, who had died in 1938, aged 69. After technical refinements, the first light ceremony was on 26 January 1949. Following additional work to perfect the optics, the telescope finally became available to researchers in 1950.

Thanks to Hale, for four decades Mount Wilson possessed the world's most powerful reflecting telescopes. Their light grasp was an unrivalled advantage in the study of faint nebulae. Even when the 200in claimed the record as the largest telescope on Earth, it was managed by the same organisation based in Pasadena.

**ABOVE** Mount Palomar in 1937 with the dome housing the 18in Schmidt survey telescope and construction of the much larger dome for the 200in telescope. *(Palomar/Caltech/E.W. Gray)*

**BELOW** Almost 1,000 invited guests in the dome on Mount Palomar attend the dedication of the 200in reflecting Hale Telescope on 3 June 1948. *(Caltech and the Palomar Observatory)*

Shapley received directorship of the Harvard College Observatory. By the early 1920s, America's first observatory had become an astronomical auditing house, compiling vast catalogues of raw data. In contrast, Mount Wilson, with the most powerful telescopes in the world, was the base camp for celestial explorers. In his memoirs, Shapley would describe his time at Harvard as "anticlimactic".

In 1917 Edwin Powell Hubble was awarded a doctorate in astronomy by the University of Chicago under the supervision of Edwin Brant Frost, director of the Yerkes Observatory. His intention was to join the Mount Wilson Observatory but America's entry into the Great War in Europe led him to sign up for the infantry. On his return in the autumn of 1919 he contacted Hale, who confirmed that the offer of a job was still open.

Hubble's first access to the 100in telescope was Christmas Eve, and in almost perfect seeing he gained plates of a number of objects to explore the capabilities of the instrument. As his thesis at Yerkes had been on faint nebulae, it was logical that he should pursue that line of research using the more powerful telescope. His plan was to start by photographing the brighter stellar nebulae in order to gain a sense of their structure and content, and then to investigate fainter ones with the aim of developing a method by which their distances could be measured. It was a bold plan.

Photographing faint nebulae required dark skies, so it couldn't be done when the Moon was bright. During his first year he was assigned 41 nights on the 100in. During that time he took 115 plates with exposures typically of 45 to 120min but in some cases as long as 9hr, which was essentially an entire night.

After inspecting all types of diffuse nebulae to assess his doctoral work with the larger telescope, in 1922 Hubble decided to study spirals, and in particular to seek novae associated with them.

On 4 October 1923, using an improved emulsion, Hubble took a 40min plate of M31 in poor seeing through the 100in that produced a possible nova. The next night he took another plate of a slightly longer exposure in better seeing that not only confirmed the suspected nova but also showed two other candidates. At that point, his time on the telescope was up. On returning to Pasadena, he examined a number of plates in the archive which had been exposed over the years by other astronomers. This established that one of his three possible novae was actually a variable star. This proved that he was resolving an individual star. What is more, by plotting its light curve in the following months he determined its period to be 31.4 days with the characteristic profile of a Cepheid. The accepted calibration put it at a distance of about 1 million light years.

In early 1924 Shapley received a letter from Hubble dated 19 February which casually said, "You will be interested to hear that I have found a Cepheid variable in the Andromeda Nebula." Hubble sent a hand-drawn light curve that "rough as it is, shows the Cepheid characteristics in an unmistakable fashion". Hubble also pointed out that "in the last five months [I] have netted nine novae and [a second] variable". The second variable was farther out in the spiral than the Cepheid and too dim (on the plates) for Hubble to accurately measure its period, but 21 days seemed reasonable, the shorter period explaining why it was fainter and harder to follow. He finished off with, "I have a feeling that more variables will be found by careful examination of long exposures. Altogether next season should be a merry one and will be met with due form and ceremony." The casual style of the letter belied the fact that Hubble and Shapley detested one another.

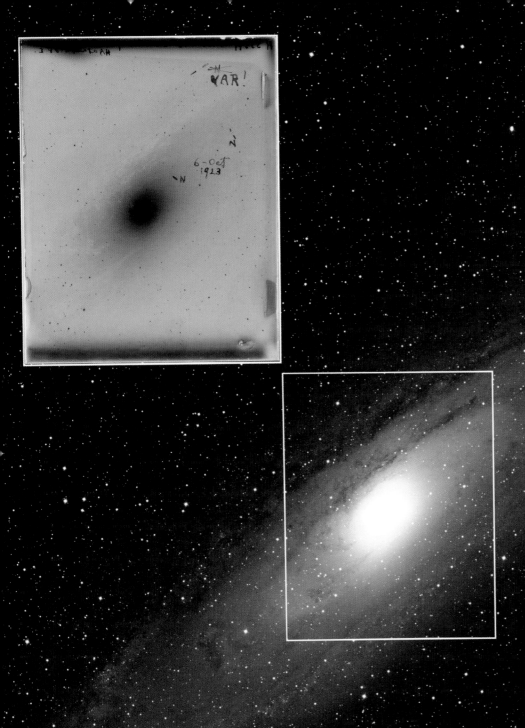

VAR!

6-Oct
1923

Showing the letter to Cecilia Helena Payne, an English-born postgrad under his tutelage, Shapley said, "Here is the letter that has destroyed my universe." At a distance of a million light years, a system of stars as large as the M31 spiral had to be independent of the Milky Way system.

By August 1924, a year into his project, Hubble had confirmed another dozen variables in M31, at least three of which were Cepheids. In addition, he had found variables of as yet unidentified types in the spiral M33 in Triangulum and some suspects in M81 and M101, both of which were in Ursa Major.

The news that M31 was almost a million light years away spread informally until it was made public on 1 January 1925 in a paper read on Hubble's behalf by Henry Norris Russell at a joint meeting of the American Astronomical Society and the American Association for the Advancement of Science in Washington DC.

Even using the 100in, Cepheids could be detected only in the nearest spirals, and Hubble was able to employ them to measure the distances to only six cases of what he decided to call 'extra-galactic nebulae' (although other astronomers used the term 'galaxy', as we shall do here). In order to probe farther, Hubble assumed that the brightest stars in any galaxy were of similar luminosity and, like Shapley before him with the globular clusters, employed these as standard candles. When he was no longer able to resolve stars, he assumed that galaxies that had similar form were of similar luminosity. Lacking evidence to the contrary, these were fair assumptions. In essence, he was following the example of William Herschel, who presumed that all stars were equally luminous to estimate their relative distances and obtain some insight into their distribution in space. Except here, Hubble was dealing with entire galaxies rather than individual stars. On

**BELOW** The light curve which Edwin Hubble manually plotted for the first Cepheid variable to be discovered in M31. On 19 February 1924 he sent it to Harlow Shapley. *(Harvard College Observatory)*

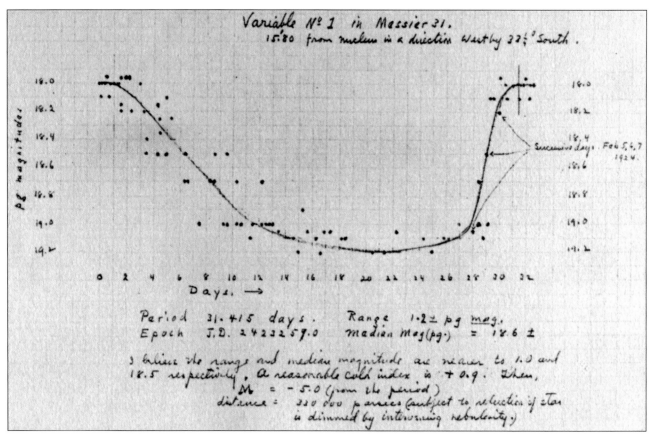

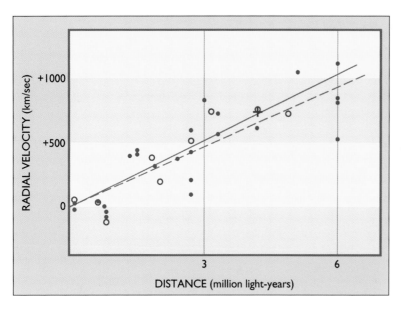

ABOVE The plot of redshift versus distance for spiral nebulae which Edwin Hubble published in *Proceedings of the National Academy of Sciences* in 1929. *(W. D. Woods)*

BELOW Milton Humason (on the left), Edwin Hubble, Walter Baade, and Rudolf Minkowski in the library of the Mount Wilson Observatory office in Pasadena. *(Mount Wilson Observatory)*

discerning a ten-fold difference in brightness between the most and the least prominent members of a cluster of galaxies in the direction of Virgo, Hubble assumed them to be equally luminous and evenly distributed within a sphere, then calculated how far away those galaxies for which he had distances would require to be in order to appear as faint as the mid-point in this brightness range. He thus estimated the centre of the cluster to lie at a distance of 6.6 million light years. (Note: As with early estimates of the distances to stars, the distances to galaxies would be revised as techniques improved.) By early 1928, Hubble had measured distances for 24 of the spirals for which Slipher had measured redshifts.

While attending the International Astronomical Union meeting hosted by the University of Leiden, Hubble met Willem de Sitter. After Hubble had outlined his work, de Sitter told him that his solution of Einstein's field equations predicted a direct correlation between distance and redshift displacement. However, as it had not specified the rate at which the displacement ought to increase with distance, he urged Hubble to measure redshifts of spirals that were too faint for Slipher to have measured. It would be interesting to determine whether such a relationship existed and, if so, to measure the slope of the plot.

In 1905, at the age of 14, Milton La Salle Humason attended a summer camp on Mount Wilson. He then dropped out of school and worked as a bellboy at the 'local hotel', which was a single-storey log structure and a number of shacks that were owned by the company which opened the perilously narrow access route to the summit of the mountain. It was very rustic, and his duties included tending to pack animals. By 1910 he was a mule driver for the pack trains that carried loads up the trail. In 1911 he married a daughter of an engineer at the observatory. They settled in Pasadena. In 1917, with the 100in telescope undergoing trials, Humason became janitor/night assistant. In 1918 he started taking plates to monitor variable stars, particularly for Shapley, who did not like the long, cold nights. Two years later, at Shapley's recommendation, Humason was accepted on to the staff of the observatory. In 1924 he was promoted to assistant astronomer.

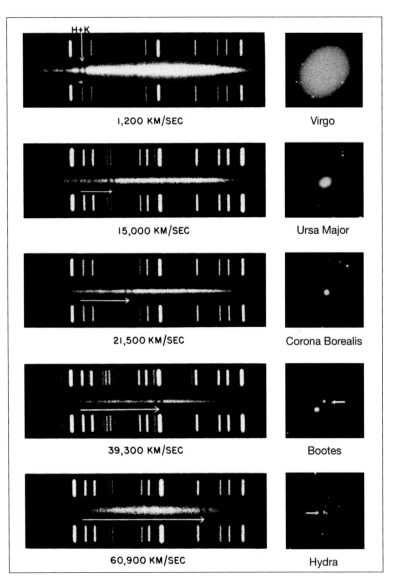

1,200 KM/SEC — Virgo

15,000 KM/SEC — Ursa Major

21,500 KM/SEC — Corona Borealis

39,300 KM/SEC — Bootes

60,900 KM/SEC — Hydra

By 1928 Humason was an expert in the techniques of astrophotography.

Upon returning from his European trip, Hubble pointed out to Humason that Slipher had hit the limit of what the Flagstaff telescope could do in measuring the redshifts of spirals, and said it would be interesting to use the 100in to determine whether the redshifts of fainter spirals increased in proportion to their distances. It was agreed that Hubble would continue to measure distances to spirals and his colleague would measure their redshifts.

For his first target, Humason chose a spiral for which Hubble had a distance and Slipher had found too faint to measure. Even with the 100in, it was a tedious exercise of integration over frigid nights for an exposure of 33hr. The spectrum was weak but measureable, and the resulting redshift of 3,000km/sec supported the relationship. Humason spent a week integrating an exposure of 45hr to obtain a superior spectrum that confirmed the velocity.

As Humason endeavoured to attain even longer exposures for fainter spirals, Hale ordered the workshop of the observatory to develop a 'faster' spectrograph and a better camera to obtain over a few hours a spectrum that would have taken several nights using the original equipment. This cut down the exposure time by concentrating the spectrum into the smallest possible length and width. The plate was just 1⁵⁄₁₆in by ⁵⁄₈in. Although tiny, the spectrum would show the absorption lines that were necessary for a redshift measurement. Armed with this tool, Humason obtained spectra of all 45 spirals which Slipher and others had measured, confirming the reported redshifts, then he started on the list of fainter spirals for which Hubble had distances.

**RIGHT** The extended plot of redshift versus distance which Edwin Hubble and Milton Humason published in the *Astrophysical Journal* in 1931. The box at the bottom left contains the results reported two years earlier. *(W. D. Woods)*

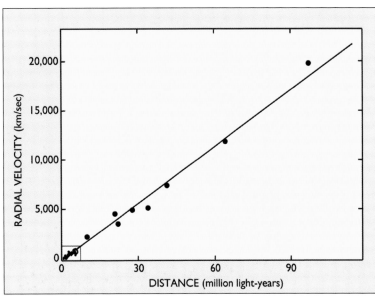

As distances and redshifts became available, Hubble plotted them on a graph and found there was indeed a relationship, in that the greater a spiral's distance, the greater its redshift. He wrote this up in the paper 'A Relation Between Distance and Radial Velocity Among Extra-Galactic Nebulae' that was published in early 1929. Hubble saw himself as an observer rather than a theoretician, so he made no attempt to explain the data. He simply pointed his readers to a paper by de Sitter, whose theoretical work had predicted such a relationship.

In the theory published by de Sitter in 1917, a redshift was not a radial velocity due to the expansion of space, it was a slowing down of time at great distances and was proportional to separation. In later years, several people showed that the 'de Sitter effect' was equivalent to a universe that was expanding. When Hubble reported that such a relationship existed, most astronomers opted to interpret the redshifts as radial velocities and hence evidence that the universe was in a state of expansion. Ever cautious, Hubble refused to commit himself to an interpretation, preferring even as late as the mid-1930s to describe the redshifts only as 'apparent radial velocities'.

On average, the velocity increased by about 160km/sec per million light years of additional distance. In more familiar terms, it was equivalent to the separation of two objects 1,000km apart increasing by a fraction of a millimetre per annum. The geometrical form of the relationship meant that, by and large, the galaxies were racing away from each other at a constant rate. It was this announcement that caused Eddington to call for a reassessment of Einstein's static model of the universe.

Einstein had put aside the cosmological implications of General Relativity by the mid-1920s in order to focus on other issues, but in early 1931, having heard of Hubble's results, he eagerly accepted an invitation to visit Mount Wilson during a sabbatical that included a visit to the California Institute of Technology located in Pasadena. He was driven in the rear of a touring car, seated between Hubble and Adams, director of the observatory. At the summit there was a welcoming party which included leading astronomers from other establishments. During a tour of the telescopes, Einstein expressed his awe at the 100in. There was nothing to rival it. That evening, he peered through the telescope at several astronomical objects, retired to a cottage on the mountain after midnight, and was up again in time for sunrise.

On 3 February, Einstein told the journalists who were accompanying him on his visit that he had abandoned his static model of the universe

because the work undertaken at Mount Wilson had demonstrated that the universe wasn't static in the sense of being unchanging with time, it was dynamic, in a state of expansion. He acknowledged that he and de Sitter had been wrong, and that Friedmann and Lemaître were right.

Rather belatedly, Einstein realised that when his field equations indicated that the universe could not be static and he had introduced the cosmological constant to prevent it from collapsing under self-gravitation, his logic had been flawed. He had failed to consider the possibility that the universe might have originated in a state of expansion. Einstein had been unlucky. His theoretical analysis was ahead of the observational evidence, and with astronomy not being his area of expertise he had tried to make his theory 'fit'. If only he had let his field equations speak for themselves, he would have predicted the state of expansion that Hubble had just detected.

Only in 1932, by which time Hubble and Humason had data that extended the relationship out to 100 million light years with a redshift of almost 20,000km/sec, did Einstein and de Sitter publish a dynamic solution to the field equations. Their joint paper fully acknowledged the work done independently by Friedmann and Lemaître.

They expressed General Relativity in its simplest form, eliminating the cosmological constant. The result was a mathematically 'flat' universe whose rate of expansion would be slowed by self-gravitation, but never halt. It explained the illusion that we were positioned at the centre of the universe by pointing out that the galaxies were not really moving through space at the speeds implied by their radial velocities, it was space that was expanding and carrying the galaxies with it. The exponential expansion was a geometrical effect. As Eddington would later observe in explaining the phenomenon, precisely the same thing occurs for points on the surface of an inflating balloon.

Although the Einstein-de Sitter solution implied (as indeed did the 1931 paper by Lemaître) that the universe would have expanded from a progenitor of infinite density, this was deemed preposterous. Nevertheless, they published the solution without thinking that it might describe reality. It was merely yet another possible solution to the field equations. In effect, of course, the Einstein-de Sitter universe was a special case of Friedmann's solutions with the cosmological constant set to zero.

On a visit to Pasadena in January 1933, Einstein met Lemaître a second time and complimented his work as "the most beautiful and satisfactory explanation of creation".

## Chapter Four

# The fireball

As physicists investigated the origin of the universe, they discovered that it must have been created in what one eminent scientist derided as a 'big bang'. However, there was observational evidence for the 'afterglow' of this fireball. Although the lightest elements, hydrogen and helium, were synthesised immediately, the heavier elements were made by nuclear fusion in the cores of stars.

**OPPOSITE** The Bell Laboratories horn antenna on Crawford Hill in Holmdel, New Jersey, which discovered the cosmic microwave background. *(W. D. Woods)*

# Primordial nucleosynthesis

**W**hile attending the University of Cambridge, England, Cecilia Helena Payne was encouraged by Arthur Stanley Eddington to pursue astrophysics. Although the university didn't allow women to graduate, on finishing in 1923 she received a fellowship to pursue a PhD at the Harvard College Observatory, where the director, Harlow Shapley, set her the challenge of analysing stellar spectra. Her thesis 'On Stellar Atmospheres' in 1925 was proclaimed as "brilliant". Against the prevailing opinion, she argued that the Sun was primarily hydrogen. In addition to concluding that the relative abundances of the elements in stars were similar to the Sun, she introduced the term 'cosmic abundances'.

George Gamow's interest in cosmology was stimulated by attending lectures by Alexander Alexandrovich Friedmann at the University

**LEFT Cecilia Helena Payne gained her doctorate on 'Stellar Atmospheres' at the Harvard College Observatory.** *(Harvard College Observatory)*

**BELOW Cecilia Payne's doctoral thesis.**

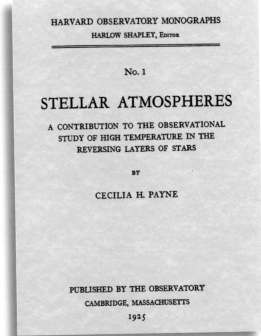

HARVARD OBSERVATORY MONOGRAPHS

HARLOW SHAPLEY, EDITOR

No. 1

## STELLAR ATMOSPHERES

A CONTRIBUTION TO THE OBSERVATIONAL STUDY OF HIGH TEMPERATURE IN THE REVERSING LAYERS OF STARS

BY

CECILIA H. PAYNE

PUBLISHED BY THE OBSERVATORY
CAMBRIDGE, MASSACHUSETTS
1925

of Leningrad in 1924 as a postgraduate. He then embarked on a tour that took him to the University of Göttingen in Germany, where he gained his doctorate in 1928 for work related to the nature of the atomic nucleus. After posts at the Institute of Theoretical Physics in Copenhagen and the University of Cambridge, England, he returned to Russia in 1931 and was so disheartened by the political situation that on another tour in 1933 he sought an appointment at George Washington University in Washington DC.

Gamow endorsed Georges Lemaître's idea that the universe exploded out of a superdense 'primeval atom'. However, the theory could not explain the fact that hydrogen was predominant. As an alternative, Gamow suggested that Lemaître's 'atom' produced only neutrons, and since free neutrons are unstable these would have progressively decayed into equal numbers of protons and electrons. Gamow therefore argued that the nuclei of the heavier elements were made by the fusion of protons with late-decaying neutrons. He announced this theory in a seminar at Ohio State University in 1935. Whereas his predecessors, most notably Einstein, had treated cosmology as a study in geometry, Gamow investigated it in terms of high-energy physics.

Gamow continued to develop this cosmological theory while consulting at the Applied Physics Laboratory of Johns Hopkins University in Baltimore during the war, then in 1946 wrote a paper provocatively entitled 'The Expanding Universe and the Origin of the Elements'.

When Ralph Asher Alpher abandoned his doctoral work on cosmic rays after finding that he was a year or so behind a rival at another institute, Gamow agreed to supervise him in a study of fusion processes, using recent experimental data by the Argonne National Laboratory in Chicago on how neutrons were captured. In his calculations Alpher verified that helium would be generated, but at that point the process stalled. Although the idea of 'primordial nucleosynthesis' was flawed, Gamow cheekily pointed out that by explaining the abundances of hydrogen and helium it successfully accounted for 99% of the matter in the universe.

What was fascinating was that the neutron capture process was potentially so efficient that it could easily have turned all the hydrogen into helium. The helium abundance (around 8% by number and 25% by mass) therefore meant there must have been an intense radiation field which acted to break up the helium, with the outcome being a balance between all these processes.

Alpher teamed up with Robert C. Herman, a postdoc from Princeton working at Johns Hopkins, and after calculating a ratio of billions of photons of energy per nucleon (the general name for protons and neutrons in the atomic nucleus) they used an early computer to extrapolate the Friedmann equations back in time. This showed that the energy density of the primordial radiation field must have initially greatly exceeded that of matter, expressed by Einstein's famous equation $E=mc^2$. Radiation density is the amount of energy in a given volume of space, and is equivalent to the temperature of the 'black body' that emits the same amount of energy with a very characteristic spectral distribution.

**BELOW** Robert Herman (left) and Ralph Alpher in a composition in which the essence of their mentor George Gamow emerges from a bottle of Ylem, the primordial stuff of the universe. The montage was created by Alpher and Herman and slipped into a slide presentation given by Gamow in 1949.

The renowned American experimental physicist Albert Abraham Michelson ventured in 1894, "It seems probable that most of the grand underlying principles have been firmly established."

Nevertheless, several puzzling observations had been made over the years. In particular, Gustav Robert Kirchhoff realised in 1860 that incandescent gases emit at the same wavelengths as they absorb when they are cold. Thus the colour of an object depends on which wavelengths of sunlight it absorbs and which it reflects. Kirchhoff postulated that a body that absorbed at all wavelengths would appear perfectly black, and once it was heated up it would radiate that same energy by emitting it across the entire spectrum.

While studying thermodynamics in 1879, Josef Stefan realised that there was a relationship between the fourth power of an object's temperature and the amount of energy it radiated. A 'black body' can therefore be defined as an object which (after being heated) emits its radiation in a spectrum that can be described solely by a characteristic temperature.

Although most bodies reflect only a fraction of the radiation that they absorb, physicists found a black body to be a mathematically useful concept.

In particular, Wilhelm Wien undertook an experiment in 1893 which enabled him to make an approximation to a black body in the shape of a cavity possessing a small hole in its wall. Any radiation entering the cavity through the hole should be absorbed and, in accordance with Kirchhoff's thinking, the radiation that later leaked out ought to have the spectrum of a black body. When Wien measured this radiation, he realised that the energy peaked at a certain wavelength and that the wavelength of the emission peak was inversely proportional to temperature. As a result, moderately hot bodies radiate primarily in the infrared, and when they are heated their peak of emission moves through the red end of the visible spectrum until they glow 'white hot'. Wien devised a mathematical formula that applied to wavelengths that were shorter than the emission peak, but which failed for longer ones.

In 1873 John William Strutt inherited the title of Lord Rayleigh, and several years later gained

**RIGHT** Wilhelm Wien.

**FAR RIGHT** John William Strutt, later Lord Rayleigh.

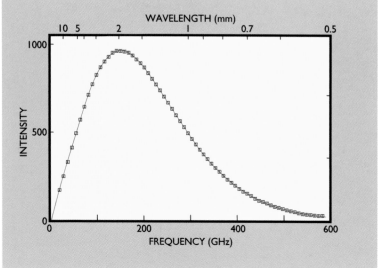

turned in favour of the Big Bang. Indeed, at an international symposium in 1973 to mark the 500th anniversary of the birth of Nicolaus Copernicus, almost every paper was in favour.

Inconclusive measurements short-ward of the turnover continued through the 1980s. It was a frustrating time for experimenters and theorists alike. Fortunately, the importance of the issue prompted the development of a satellite to provide a definitive answer.

In November 1989 NASA launched the Cosmic Background Explorer (COBE) to make measurements at 67 wavelengths in the range 0.5 to 10mm, spanning the presumed turnover. The data from the first 10min was sufficient to confirm that it was a black body spectrum. The error bars on the individual measurements were so small that the points precisely traced the predicted curve.

When John Cromwell Mather of Berkeley presented the preliminary data to a gathering of the American Astronomical Society in Arlington, Virginia, in January 1990, the audience of 2,000 scientists gave him a standing ovation. The final report put the temperature of the black body at 2.726(±0.01)K.

After several years, and many millions of individual measurements for an all-sky temperature map, COBE revealed the background to be isotropic to 1 part in 100,000. In fact, it was so smooth that the maximum departure from the mean was a mere 30-millionth of a degree. If the distribution of matter was perfectly smooth when the

primordial radiation field 'decoupled', the interaction of photons with matter would have been homogeneous and left a completely uniform field. COBE therefore revealed that there was already a variation in density at that time. It was from such 'wrinkles' that the largest-scale structures that we see today developed.

## Making elements in stars

A major mystery of the 19th century was what made the Sun shine and how long it had been doing so.

In 1854 the German physicist Hermann Ludwig Ferdinand von Helmholtz pointed out that if the Sun was in a state of slow but progressive collapse, it might be powered by the release of gravitational energy. As long as the interior of a star remained gaseous, it would contract. When the core was compressed into a liquid state it would resist further compression and cease to release gravitational energy. Thereafter, the star would rapidly radiate away its heat and cool down. William Thomson (subsequently Lord Kelvin) calculated that for the Sun to have shone for 50 million years by the release of gravitational energy its original diameter would have to have been as broad as Earth's orbit, with the implication that both Venus and Mercury must have been orbiting inside the Sun! By 1897, Kelvin was of the opinion that Earth was about 20 million years old. But geological evidence clearly indicated Earth to be considerably older. Just as there appeared to be an impasse between physics and geology, an unexpected discovery offered the prospect of a solution.

In 1896, while investigating the X-rays that had been discovered by Wilhelm Konrad Röntgen the previous year, Antoine Henri Becquerel in Paris realised that they were released by certain kinds of materials. In 1898 Marie Curie named this phenomenon radioactivity. For a while it appeared the Sun might be powered by the energy produced by the decay of heavy elements, but in 1925 Cecilia Helena Payne at the Harvard College Observatory realised that the Sun is primarily made of hydrogen. On hearing this, Eddington modelled a star as a sphere of gaseous hydrogen to determine whether the energy that held the Sun against gravitational collapse was derived from fusing hydrogen to helium. Unlike radioactive decay, the nuclear fusion process is unable to happen spontaneously; it requires a very high temperature to smash nuclei together with sufficient energy to provoke the reaction. In 1926, he estimated the temperature at the core of the Sun to be

about 15 million degrees. That seemed to be hot enough for protons (hydrogen nuclei) to overcome the mutual repulsion of their positive electric charges, but smashing together a pair of protons is only the start in the process because a stable helium nucleus also contains two neutrons.

In 1937 George Gamow and Carl Friedrich von Weizsäcker proposed that a pair of protons could create a heavy form of hydrogen called deuterium whose nucleus comprised a proton and a neutron, with the surplus positive charge being carried away by an 'antielectron'. The next year, Hans Albrecht Bethe, having fled Germany to America and taken a post at Cornell University, worked out the rest of the process. In the second step, a proton interacted with a deuterium nucleus to make a lightweight form of the helium nucleus with two protons and a neutron. As this nucleus was unstable, the process required two of these nuclei to fuse to produce the stable form with two protons and two neutrons, ejecting the surplus protons to engage in future reactions. But the process essentially halted at helium, and therefore could not account for heavier elements.

It was eventually determined that in fusing hydrogen into helium, the Sun is converting 4 million tonnes of material directly into energy every second in line with Einstein's $E=mc^2$. Despite this prodigious loss of mass, the Sun has been on the 'main sequence' (burning hydrogen) for 5 billion years and is barely halfway through this phase of its evolution.

While it is on the main sequence, the Sun will accumulate helium in its core. When the core reaches a sufficient mass, it will suddenly collapse and the release of gravitational energy will blow off the outer atmosphere of the Sun. At the same time, the core will be heated sufficiently to initiate helium fusion, causing the Sun to leave the main sequence and become a 'red giant' star.

In 1945, Hoyle at Cambridge decided to figure out whether elements heavier than helium were created in stars. He published his results in two papers in 1946 and 1947. William Alfred Fowler at the California Institute of Technology decided to measure the reaction rates in order to refine Hoyle's theory.

Bethe had also studied a process for fusing hydrogen into helium in which a carbon nucleus

**BELOW** The proton-proton chain that converts hydrogen to helium.
*(W. D. Woods)*

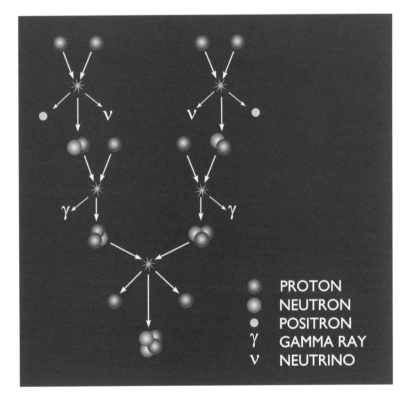

PROTON
NEUTRON
POSITRON
γ GAMMA RAY
ν NEUTRINO

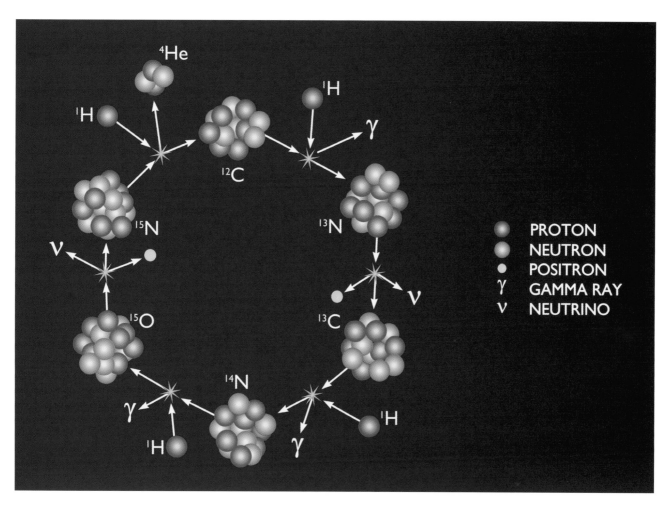

PROTON
NEUTRON
POSITRON
$\gamma$ GAMMA RAY
$\nu$ NEUTRINO

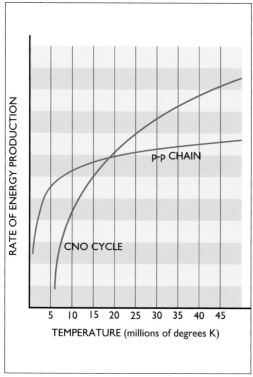

**ABOVE** The carbon-nitrogen-oxygen cycle that converts hydrogen to helium. *(W. D. Woods)*

**FAR LEFT** The efficiencies of the proton-proton and carbon-nitrogen-oxygen fusion processes. For a star with a core temperature exceeding 18 million degrees the CNO cycle is dominant. *(W. D. Woods)*

**LEFT** Edwin Ernest Salpeter.

ABOVE **Margaret and Geoffrey Burbidge with William Fowler (right) upon the latter's 60th birthday in July 1971 when, being a lifelong stream-train enthusiast, he was presented with a working model engine.**

absorbed a succession of protons, growing progressively heavier until it spontaneously decayed back to its original form by spitting out a helium nucleus. With the 'proton-proton' process stalled at helium, it was not known where the carbon for this catalytic process could have come from, because if two helium nuclei were to fuse to produce beryllium-8 this would promptly decay. In 1952 Edwin Ernest Salpeter at Cornell realised that if this were to take place in a helium-rich stellar core then there was a chance the short-lived beryllium would fuse with a third helium nucleus to make a stable carbon-12 nucleus.

To wrap up the story, in 1957 Fowler, Hoyle, and the husband and wife team of Geoffrey and Margaret Burbidge at Cambridge published a seminal paper that explained how elements heavier than helium are manufactured in very hot stellar cores.

Interestingly, stellar nucleosynthesis couldn't account for the observed helium abundance. It had been explained by Gamow, but his primordial nucleosynthesis proposal had fallen on stony ground. All that was required to fully understand cosmic abundances was for someone to 'join the dots'.

In the early 1960s, Hoyle and Roger John Tayler calculated how much helium would have been made by primordial nucleosynthesis in a hot fireball, finding it would have produced at least 14% helium in terms of mass. Hoyle's collaborator, William Fowler, then asked Robert V. Wagoner, a Stanford physicist who had just joined his team, to calculate the heavier elements that would emerge from such a fireball. In 1966 Wagoner confirmed that the process would stall at helium with an abundance of about 25%; this being the observed value.

It is ironic, therefore, that Hoyle, one of the most vocal critics of a primordial fireball, should have been one of the authors of the research that convinced most cosmologists to take the idea seriously!

## The Periodic Table of Elements

| 1A | 2A | 3B | 4B | 5B | 6B | 7B | 8B | 8B | 8B | 1B | 2B | 3A | 4A | 5A | 6A | 7A | 8A |
|---|---|---|---|---|---|---|---|---|---|---|---|---|---|---|---|---|---|
| 1 H 1.00794 | | | | | | | | | | | | | | | | | 2 He 4.002602 |
| 3 Li 6.941 | 4 Be 9.012182 | | | | | | | | | | | 5 B 10.811 | 6 C 12.0107 | 7 N 14.0067 | 8 O 15.9994 | 9 F 18.9984032 | 10 Ne 20.1797 |
| 11 Na 22.989769 | 12 Mg 24.3050 | | | | | | | | | | | 13 Al 26.9815386 | 14 Si 28.0855 | 15 P 30.973762 | 16 S 32.065 | 17 Cl 35.453 | 18 Ar 39.948 |
| 19 K 39.0983 | 20 Ca 40.078 | 21 Sc 44.955912 | 22 Ti 47.867 | 23 V 50.9415 | 24 Cr 51.9961 | 25 Mn 54.938045 | 26 Fe 55.845 | 27 Co 58.933195 | 28 Ni 58.6934 | 29 Cu 63.546 | 30 Zn 65.38 | 31 Ga 69.723 | 32 Ge 72.64 | 33 As 74.92160 | 34 Se 78.96 | 35 Br 79.904 | 36 Kr 83.798 |
| 37 Rb 85.4678 | 38 Sr 87.62 | 39 Y 88.90585 | 40 Zr 91.224 | 41 Nb 92.90638 | 42 Mo 95.96 | 43 Tc [98] | 44 Ru 101.07 | 45 Rh 102.90550 | 46 Pd 106.42 | 47 Ag 107.8682 | 48 Cd 112.411 | 49 In 114.818 | 50 Sn 118.710 | 51 Sb 121.760 | 52 Te 127.60 | 53 I 126.90447 | 54 Xe 131.293 |
| 55 Cs 132.9054519 | 56 Ba 137.327 | 57-71 Lanthanides | 72 Hf 178.49 | 73 Ta 180.94788 | 74 W 183.84 | 75 Re 186.207 | 76 Os 190.23 | 77 Ir 192.217 | 78 Pt 195.084 | 79 Au 196.966569 | 80 Hg 200.59 | 81 Tl 204.3833 | 82 Pb 207.2 | 83 Bi 208.98040 | 84 Po [209] | 85 At [210] | 86 Rn [222] |
| 87 Fr [223] | 88 Ra [226] | 89-103 Actinides | 104 Rf [267] | 105 Db [268] | 106 Sg [271] | 107 Bh [272] | 108 Hs [270] | 109 Mt [276] | 110 Ds [281] | 111 Rg [280] | 112 Cn [285] | 113 Uut [284] | 114 Fl [289] | 115 Uup [288] | 116 Lv [293] | 117 Uus [294] | 118 Uuo [294] |

**Lanthanides**

| 57 La 138.90547 | 58 Ce 140.116 | 59 Pr 140.90765 | 60 Nd 144.242 | 61 Pm [145] | 62 Sm 150.36 | 63 Eu 151.964 | 64 Gd 157.25 | 65 Tb 158.92535 | 66 Dy 162.500 | 67 Ho 164.93032 | 68 Er 167.259 | 69 Tm 168.93421 | 70 Yb 173.054 | 71 Lu 174.9668 |
|---|---|---|---|---|---|---|---|---|---|---|---|---|---|---|

**Actinides**

| 89 Ac [227] | 90 Th 232.03806 | 91 Pa 231.03588 | 92 U 238.02891 | 93 Np [237] | 94 Pu [244] | 95 Am [243] | 96 Cm [247] | 97 Bk [247] | 98 Cf [251] | 99 Es [252] | 100 Fm [257] | 101 Md [258] | 102 No [259] | 103 Lr [262] |
|---|---|---|---|---|---|---|---|---|---|---|---|---|---|---|

Legend: Alkali Metals | Alkaline Earth | Transition Metal | Basic Metal | Semi Metal | Non Metal | Halogen | Noble Gas | Lanthanides | Actinides

**ABOVE** The Periodic Table of Elements. *(Todd Helmenstine)*

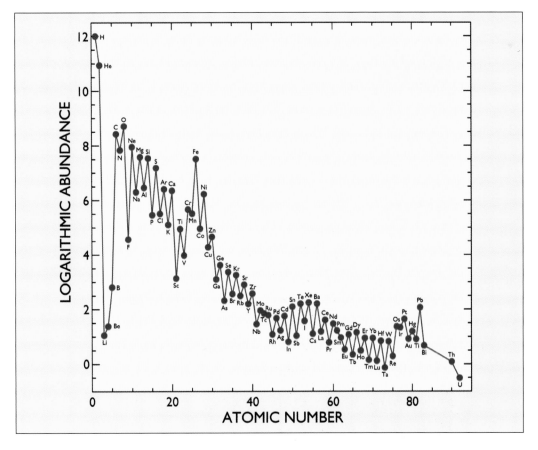

**LEFT** Solar relative elemental abundances. *(Extended by Harland from a plot by M. Asplund et al., in 'The Chemical Composition of the Sun', Annual Review of Astronomy and Astrophysics vol.47, pp481-522, 2009/ artwork W. D. Woods)*

## Chapter Five

# Black holes

General Relativity predicted that if an object were sufficiently dense, light would not be able to escape. When astronomers studied celestial sources of X-rays they discovered that 'black holes' really do exist. Studies of radio galaxies and quasars revealed that there are supermassive black holes in the cores of galaxies, including our own Milky Way system.

**OPPOSITE** A depiction of a stellar mass black hole in a binary star system, with gas drawn from the star spiralling inward through an accretion disc into the hole. The strongly wound magnetic field produces the polar jets of energetic plasma. *(NASA/CXC/M. Weiss)*

# Quasars

On being hired by the Bell Telephone Company at Holmdel in New Jersey in 1928, Karl Guthe Jansky was assigned the job of identifying the various sources of 'noise' that were impairing the recently introduced transatlantic radio-telephone service. He erected a large rotating antenna on nearby Crawford Hill and set to work. Having eliminated all known sources, he concluded in 1931 that there was a residual 'hiss' from the sky. By 1932 he had identified the source as lying in the direction of the centre of the galaxy.

Although Jansky published his results in 1933 in the *Bell Systems Technical Journal*, which few, if any, astronomers would have

seen, the *New York Times* ran the finding as its lead story on 5 May 1933: 'New radio waves traced to the centre of the Milky Way'.

In 1937 Grote Reber, a radio engineer in Wheaton, Illinois, built a dish antenna on a tiltable mount in his back yard to chart the radio sky at a wavelength of 2m. In 1938 he took his map to the nearby Yerkes Observatory, where Otto Struve, the director, realised not only that there was strong emission in the plane of the Milky Way but also that the dust which impeded the view at optical wavelengths was transparent to radio, enabling Reber to detect sources at the galactic centre itself.

James Stanley Hey at the Royal Radar Establishment in England inspected a strong radio source in 1945 that Reber had charted in

**BELOW Karl Guthe Jansky with his 'merry go round' antenna circa 1933 and signal processing apparatus.** *(NRAO-AUI)*

ABOVE **Grote Reber.** *(NRAO/AUI)*

RIGHT The tilting antenna built by Grote Reber at his home in Wheaton, a suburb of Chicago, Illinois. He would set it at a given elevation and record the signal over a period of 24hr. By adjusting the angle, in a couple of weeks he was able to assemble a map of the sky from his location. *(NRAO/AUI)*

BELOW The pioneering radio map by Grote Reber at a wavelength of 1.9m that was published in the *Astrophysical Journal* in 1944. There was a clear match with the plane of the Milky Way and concentrations in the constellations Cassiopeia, Cygnus and Sagittarius. *(NRAO/AUI/W. D. Woods)*

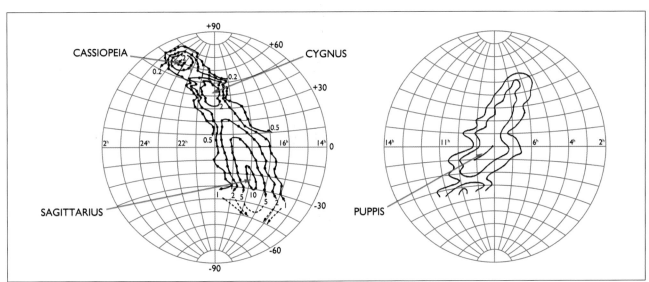

**LEFT** The 200in Hale telescope on Mount Palomar. *(PalomarSkies)*

Cygnus, prompting Hey to designate it Cygnus A. The fact that its intensity was fluctuating on a timescale of minutes meant the source could be no larger than the distance that light was able to travel in that interval.

John Gatenby Bolton was too far south to observe Cygnus A, but in 1947 he built an antenna atop Dover Heights, a cliff near Sydney, Australia, in order that a point-like source rising over the horizon could be detected both directly in the sky and by its reflection off the sea. This interferometry technique yielded much finer angular resolution than his antenna could offer alone. In this manner, Bolton was able to associate the Virgo A radio source with the giant elliptical galaxy M87.

Francis Graham Smith at the University of Cambridge localised the position of Cygnus A sufficiently in 1951 to facilitate an optical search. Smith airmailed the position to Walter Baade at the California Institute of Technology, who inspected it with the 200in telescope on Mount Palomar and found a rich cluster of galaxies, at the centre of which was a peculiar 18th magnitude object that appeared to have a distorted wispy structure and a double nucleus.

Having recently co-authored a paper speculating on the collision of galaxies, Baade concluded that this was such an event. A spectrum by his colleague Rudolf Leo Bernhard Minkowski showed strong emission lines, indicating a hot plasma. The fact that the radial velocities of the two components were identical ruled out the colliding-galaxies scenario. Although Cygnus A was one of the brightest radio sources in the sky, the Doppler redshift implied it was almost 1 billion light years distant. The mystery was how such a remote object could generate such a strong radio signal.

In 1953 Roger Clifton Jennison in England discovered that the discrete source was

**LEFT** The double-node of the Cygnus A radio source revealed in a 'deep image' obtained by Walter Baade using the 200in telescope in 1951 was something entirely new.

Top: A map made in 1953 by R. C. Jennison revealed the 'discrete' radio source Cygnus A to sit between two lobes. Middle: As radio telescopes improved, the lobes were investigated in great detail, as shown by the three maps by the Very Large Array in Socorro, New Mexico. Bottom: With improved computers it was possible to turn radio maps into beautiful false-colour images, as shown here for a wavelength of 6cm. *(NRAO/AUI/R. Perley, C. Carilli and J. Dreher)*

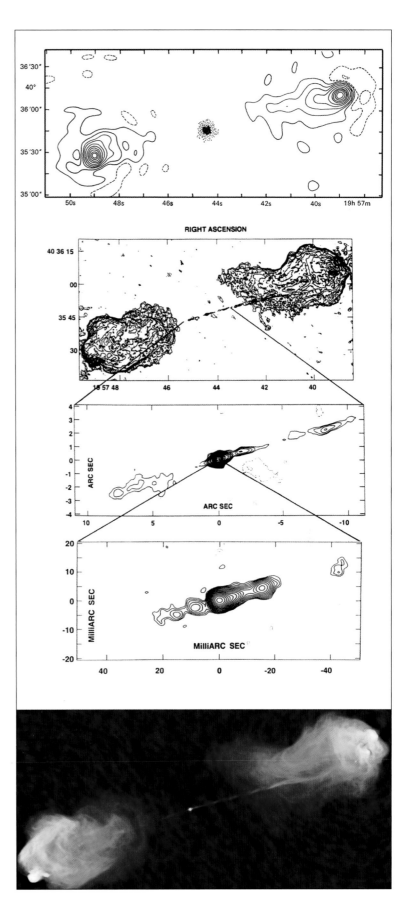

situated between two lobes of radio emission which formed a vast dumb-bell shape. Whereas the discrete source was a mere 10 seconds of arc across, the lobes spanned 3 minutes of arc. Something in this galaxy was firing material 1 million light years into space. Furthermore, that process must have been operating for a considerable time. Geoffrey Burbidge, then at Cambridge, suggested in 1958 that the radio emission indicated a process such as synchrotron radiation by electrons spiralling in an intense magnetic field. This is known as non-thermal radiation to distinguish it from thermal emission from a 'black body' at a specific temperature. Non-thermal emission meant there must be a 'powerhouse' operating in the core of that galaxy.

After using an interferometer array to survey the sky, Martin Ryle's Cambridge group published a series of catalogues. In their third, in 1958, the sources had the prefix '3C' and were ordered in terms of their right ascension on the sky.

Although some radio sources correlated with galaxies, others were difficult to identify owing to the substantial 'error box' for their radio position. However, in cases where the Moon travelled in front of a discrete radio source it was possible to time when the source was occulted by the Moon and, using data from multiple occultations, refine the position sufficiently to facilitate an optical search.

Cyril Hazard monitored 3C273 in Virgo when it was occulted by the Moon in 1962 using the 64m dish of the Parkes telescope in New South Wales, Australia. At the moment of occultation, the target happened to be very close to the horizon, requiring some trees to be trimmed to give the telescope a clear line of sight. The

results revealed 3C273 to be a pair of sources separated by some 20 seconds of arc. When Allan Rex Sandage at the California Institute of Technology had inspected the Cambridge position for 3C273 at Mount Palomar, the only unusual object was a 13th-magnitude star with a long, narrow jet. It was now realised that one of the radio sources coincided with the star and

the other with the tip of the jet. But what kind of star could produce such a jet?

Sandage's colleague Maarten Schmidt took a spectrum of 3C273 in December 1962 that showed several spectral lines whose positions were puzzling. However, in an epiphany on 5 February 1963 he noticed the lines were correctly spaced for hydrogen, but displaced

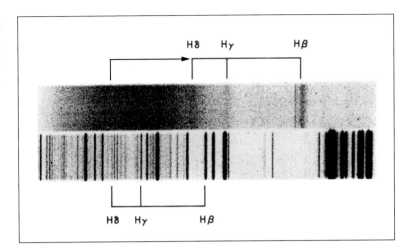

**RIGHT** It was when analysing this spectrum of 3C273 that Maarten Schmidt realised that what seemed to be a local star had a redshift implying it must be farther away than many galaxies.

by a Doppler redshift that implied a recession velocity of 47,000km/sec which, if this was interpreted in terms of Edwin Hubble's redshift-distance relationship, put the object several billion light years away. If it were that far away, then its jet had to be enormously long. In order for 3C273 to appear as a 13th-magnitude star, it had to be 100 times more luminous than any of the known radio galaxies, the most powerful of which were the dominant members of their clusters at optical wavelengths. 3C273 thereby became the first quasi-stellar radio source, soon shortened to 'quasar', and its discovery overturned the conventional wisdom about the origin and evolution of the universe.

One hundred and fifty quasars had been found by the end of the 1960s, and of the two-thirds whose spectra had been obtained, all were more heavily redshifted than 3C273, which turned out to be the nearest! Nevertheless, it was more remote than most galaxies known at that time. It was evident that quasars must be a type of galaxy that had an ultra-luminous core, but it was many years before improved telescopes, instruments, and methods enabled us to glimpse the structure of even the nearest ones.

The discovery by Arno Allan Penzias and Robert Woodrow Wilson at Holmdel of the cosmic microwave background from the primordial fireball soon led to the development of the 'standard model' of the Big Bang.

Although this theory predicted a 'cut-off' in the population of galaxies at a high redshift, representing an early point in the history of the universe, this transition was beyond the most distant measured galaxies and hence not evident. However, quasars offered a means of probing to greater redshifts.

By 1974, with only the barest of hints of a cut-off in the 200 redshifts that were available, Maarten Schmidt and Richard Green fitted the 18in Schmidt on Mount Palomar with a filter to highlight intensely blue star-like objects and exploited the telescope's wide field of view

to make a survey that discovered dozens of quasars which supported the case for a cut-off.

Then Frank James Low and Harold Johnson created an instrument that could detect objects whose peak visible light had been displaced into the infrared. They found there were no 'high redshift' quasars in our immediate neighbourhood and only a few situated within

**BELOW** The space density of quasars over time. Surveys showed an increase in the quasar population with increasing redshifts until the so-called 'redshift cut-off' at about Z = 2.5, corresponding to a time when the universe was only about one-third of its current age. Quasars were much more common in the early universe. *(Data from Peter Shaver of the European Southern Observatory/graphic W. D. Woods)*

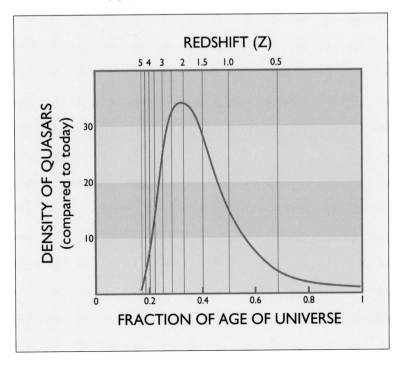

several billion light years of us. Farther out, however, the population progressively increased 1,000-fold until, at the greatest redshifts, it rapidly declined.

The quasar redshifts were evidence of an evolutionary trend in which quasars developed early in the history of the universe and were common, then somehow evolved in such a manner as to fall dormant.

## Seeking the earliest galaxies

In 1994, after astronauts had 'corrected' the optical flaw in the Hubble Space Telescope, it was able to obtain very long exposures of extremely remote galaxies. In order to study the morphology of early galaxies, three clusters were selected at a range of distances.

In the 'near' cluster, some 5 billion light years away, it found a surprisingly high percentage of spirals, many of which were interacting, and a large number of 'fragments' that seemed to have been ripped off in collisions. This suggested that interactions were commonplace when galaxies were densely packed in the early universe. Although the fragments were small, shockwaves propagating through the hydrogen clouds caused intense bursts of star formation.

The only recognisable galaxies in the 'middle distance' cluster about 9 billion light years away were of the elliptical type, which was a surprise. But there were also many fragments described as 'blue dwarfs' because they radiated intensely in the blue despite their light being heavily redshifted.

There were only a few ellipticals in the cluster at a distance of 12 billion light years away, but it also contained a quasar.

To follow up, in 1995 the Hubble Space Telescope took a 'deep field' image in which it integrated a tiny patch in Ursa Major for 120hr

**BELOW** The Hubble Space Telescope in orbit with its aperture open. It was named in honour of Edwin Powell Hubble. *(NASA)*

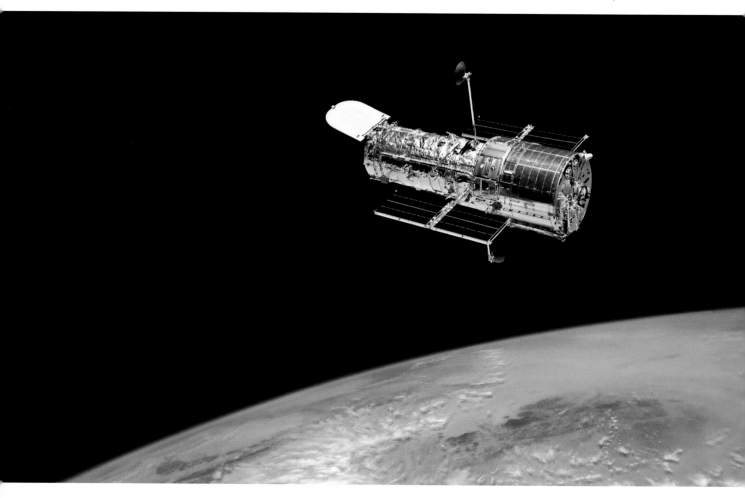

## GALAXY TYPES

Galaxies vary greatly in both shape and size. In addition to stars, spirals are usually rich in gas and dust. Pure spirals have their arms emerging directly from the nucleus, but in a 'barred' spiral the arms are connected to the ends of a linear 'bar' which spans the nucleus. Elliptical galaxies have little internal structure, are composed predominantly of very old stars, and are clear of gas and dust. They are either spherical or resemble flattened spheres. And then there are irregulars that lack a definite shape but are rich in gas; the Magellanic Clouds are irregulars.

About 75% of all galaxies are spirals (of which about two-thirds have central bars). Giant ellipticals account for 20%. The remainder are irregulars. In the early universe, however, irregulars were the majority. Their numbers decreased as they were 'gobbled up' by collisions that assembled larger galaxies by accretion. There are also a large number of dwarf galaxies. Nowadays, these tend to be satellites of larger galaxies. They are of interest because, unlike the products of accretion, they are pristine.

In a spiral galaxy, the density of matter is almost constant across the disc but density waves propagate and the concentrations trigger the collapse of gas clouds to create young, hot stars. In studying M31, Walter Baade at the Mount Wilson Observatory realised in 1947 that the spiral pattern that attracts the eye is merely "a flashy and rather inconsequential adornment"; the real essence of such a galaxy is the spherical distribution of older stars.

When spiral galaxies collide, their discs are disrupted and the merged cores create a nascent elliptical galaxy. The larger an elliptical becomes, the more it will draw in companions. The giant ellipticals in the centres of large clusters can have masses as great as a trillion times that of the Sun.

The smallest dwarf galaxies are only about 10 million solar masses. Although this is similar to the mass of the largest globular clusters found in galaxies, they differ in the sense that a globular cluster is a densely packed sphere.

## GALAXY CLUSTERS

Galaxy clusters are nowadays classified as being 'regular' and 'irregular'. The regular ones are roughly spherical aggregations that are dominated by ellipticals. They are rich and dense. Irregular clusters have no discernible shape and they can contain any type of galaxy. They are clumpy, as if comprised of groups of smaller clusters.

(in fact, it consisted of 342 exposures obtained over a period of ten days). This area had three virtues. Firstly, it was out of the galactic plane and hence clear of the obscuration that limits our view near the plane of the Milky Way. Secondly, it was uncluttered by stars. And crucially, it didn't overlap any known cluster. The objective was simply to peer as far into space as possible to find out what the faintest, most remote, and therefore earliest galaxies looked like.

**BELOW The Very Large Telescope of the European Southern Observatory at Paranal in Chile comprises four large instruments and a number of smaller auxiliary ones.**
*(ESO/J. L. Dauvergne and G. Hüdepohl)*

# THE LOCAL GROUP AND THE FATE OF OUR GALAXY

In 1936 Edwin Hubble coined the term 'local group' for a dozen or so objects that he referred to as 'extragalactic nebulae' and we now call galaxies. It contains no giant ellipticals. The three largest members are spirals, and in decreasing order of size are M31 in Andromeda, our Milky Way system, and M33 in Triangulum.

Although the group is now known to have at least 50 members, most of them are dwarfs, divided almost evenly into dwarf irregulars and dwarf ellipticals, and they are bound to the spirals. The 'satellites' of the Milky Way system include the Magellanic Clouds.

The group has an overall diameter of 10 million light years with a dumb-bell distribution, and the gravitational centre lies somewhere between our galaxy and M31. The local group is in turn bound to the Virgo cluster, so is an outlier of that vast assemblage.

A dynamic analysis suggests

**OPPOSITE PAGE**

The M31 spiral in Andromeda is on a collision course with our own Milky Way galaxy. As we see it today (top) and impressions of how it will appear in 2 billion (middle) and 3.75 billion years from now (bottom). *(NASA/ESA/ STScI/Z. Levay, R. van der Marel, T. Hallas and A. Mellinger)*

**THIS PAGE** As M31 collides with our galaxy, the gravitational interaction will disrupt both spirals, hurling streamers into intergalactic space. The top and middle impressions show this 4 billion and 5 billion years from now. In 7 billion years' time, the result of the merger will be a giant elliptical galaxy (bottom). *(NASA/ESA/ STScI/Z. Levay, R. van der Marel, T. Hallas and A. Mellinger)*

that in several
billion years our
galaxy and M31 will
collide. As they pass
through one another,
stellar collisions
are vanishingly
unlikely but the tidal
distortions will disrupt
both discs, causing a
vigorous burst of star
formation and slinging
long streamers of
stars into intergalactic
space.

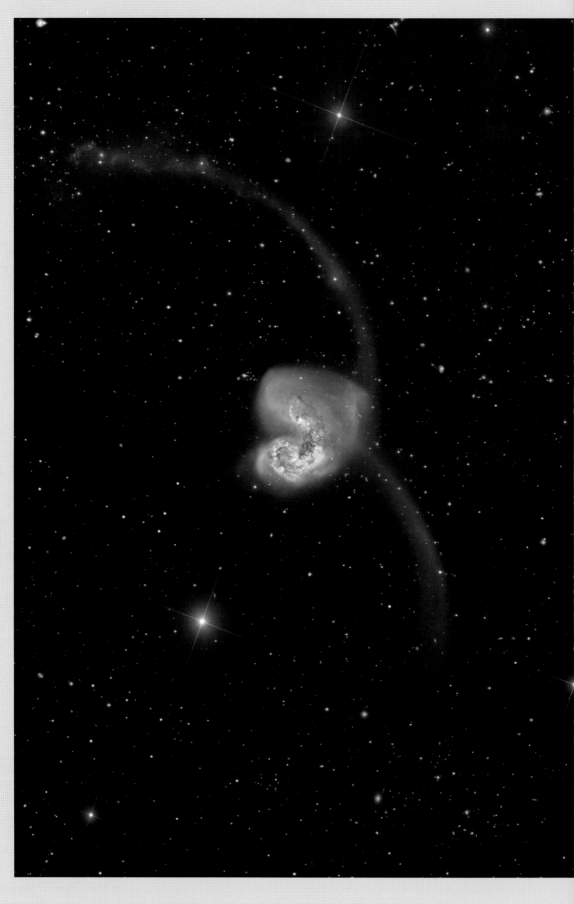

**RIGHT The
gravitational
interaction between
NGC 4038 and NGC
4039 has disrupted
the spirals and hurled
out streams of stars to
produce the Antenna
Nebula. In addition
colliding gas clouds
have given rise to
intense star formation.
Something very similar
will occur when M31
collides with the Milky
Way galaxy, several
billion years from now.
A close-up of these
interacting galaxies is
shown on the cover
of this book.** *(Subaru/
NAOJ/NASA/ESA/
STScI/R. W. Olsen,
Federico Pelliccia and
Rolf Wahl Olsen)*

The result was an astonishingly rich field of 2,000 objects, many of which were irregular 'clumps' of stars. These proved to be gravitationally bound into groups which would later coalesce as galaxies.

In 1998 the Hubble Space Telescope obtained a second deep field image on the opposite side of the sky, in the constellation of Tucana, with similar results.

During a Faint Infrared Extragalactic Survey, one of the 8.2m elements of the Very Large Telescope operated by the European Southern Observatory on Cerro Paranal in the Atacama Desert of Chile integrated for 100hr in 2002 to obtain near-infrared images of the Tucana field at a wavelength of 2.3μm, which was a part of the spectrum at which the Hubble Space Telescope was uncompetitive. The scene portrayed the situation when the universe was less than 2 billion years old. It was interesting to see that some of the early galaxies displayed spiral structure.

It was apparent that high resolution near-infrared imaging would be essential for investigating the earliest proto-galactic objects.

It will be the primary task of the long-delayed successor to the Hubble Space Telescope, the James Webb Space Telescope being built by NASA in collaboration with the European and Canadian space agencies and currently scheduled for launch in the early 2020s.

## Black holes in theory

The interior of a 'main sequence' star represents a balance in which the thermal pressure of the gas resists gravitational contraction. In the case of a white dwarf, a star that has reached the final stage of its evolution and shrunk to a radius similar to that of Earth but has a very hot surface, the interior is so compressed that the ordinary gas laws do not apply, it is a 'degenerate' gas, and gravitational collapse is staved off by the pressure of the electrons.

When Subrahmanyan Chandrasekhar found in 1930 that there was a maximum mass for a white dwarf star of roughly 1.4 solar masses, he presumed that a larger star would suffer runaway collapse.

**ABOVE** One of the 8.2m units of the Very Large Telescope of the European Southern Observatory at Paranal in Chile during 'first light' testing in 1998. *(ESO)*

Walter Baade and Fritz Zwicky at the Mount Wilson Observatory speculated in 1934 that if a star that was too massive to become a white dwarf were to collapse, the electrons would be forced into atomic nuclei, where they would combine with protons to form a 'neutron gas'. In 1939 Julius Robert Oppenheimer and graduate student George Michael Volkoff calculated how this pressure would resist further collapse and form a 'neutron star'. Hartland S. Snyder, another of Oppenheimer's students, found that if this star exceeded about 3 solar masses, the pressure of the neutron gas would be insufficient

to prevent Chandrasekhar's runaway collapse. But what would be the result?

As Albert Einstein and Willem de Sitter explored the cosmological implications of General Relativity, Karl Schwarzschild in Russia considered the prediction that the straight-line path of light would bend as it passed near a gravitating mass. He realised that if a star had a critical ratio of mass to radius, it would 'warp' space so tightly that not even a ray of light would be able to escape. This became known as the 'Schwarzschild radius'.

The mathematics of such objects was eagerly investigated, but it was presumed that in reality the process of collapse would be disrupted by turbulence and that instead of a 'singularity' forming there would be an irregular explosion. However, in 1965 Roger Penrose in London established that a runaway collapse must end in a singularity.

In 1964 Edwin Ernest Salpeter at Cornell realised that the angular momentum of material attracted by a singularity's gravity would cause it to spiral inwards to produce an 'accretion disc' in which the intense frictional heating would radiate X-rays. After discovering this independently, in 1965 Yakov Borisovich Zel'dovich and Oktay H. Gusneyov in Russia noted that if one star in a close binary system were to collapse to form a singularity, then the X-rays from the gas that it drew off its companion ought to be detectable.

At a New York conference in 1967, John Archibald Wheeler of Princeton coined the term 'black hole' to signify that a singularity constituted an actual 'hole' in the fabric of the space-time of General Relativity. The critical radius was described as the 'event horizon' in order to indicate that no events that occurred inside would be visible from outside.

## The first identified black hole

It is not possible to perform X-ray astronomy from the ground because the atmosphere is opaque in that part of the electromagnetic spectrum. The first step was therefore to place instruments on sounding rockets launched to high altitude to observe the sky for several minutes prior to falling back to Earth.

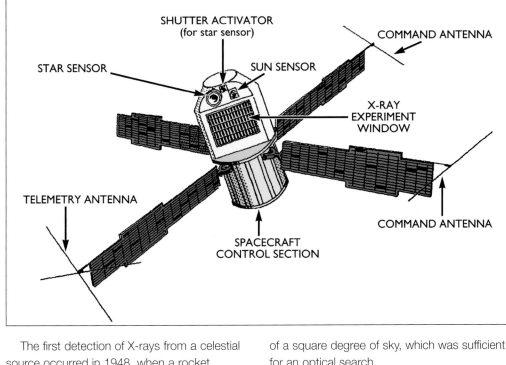

**LEFT** The Uhuru
**X-ray satellite.**
*(NASA/W. D. Woods)*

The first detection of X-rays from a celestial source occurred in 1948, when a rocket from the White Sands Missile Range in New Mexico carried an instrument created by the Naval Research Laboratory. The emission was found to be coming from the Sun. The first definite evidence of X-rays from beyond the solar system came in 1962 from a rocket carrying an instrument developed by the American Science and Engineering group, led by Riccardo Giacconi. Their plan was to find out if the Moon was fluorescing at X-ray wavelengths as a consequence of being irradiated by energetic solar wind particles, as this would allow them to study the elemental composition of the lunar surface. In fact, the Moon was a ruse to obtain funding because the government was keen to investigate the nature of its surface and indifferent to proposals to determine whether celestial sources were emitting X-rays. When the rocket went 'off course' and its instrument scanned another part of the sky this was no disappointment to the scientists, especially when the data revealed a strong source. The 100° field of view could not identify its position, but in 1963 another Naval Research Laboratory instrument with a 10° field localised it to the constellation of Scorpius, prompting the designation Scorpius X-1. By 1966 the position had been refined to within a 1,000th of a square degree of sky, which was sufficient for an optical search.

Attention soon settled on the variable star V818 Scorpii. The X-rays displayed a regular variation that had a period of 18.9hr. The blue-violet optical counterpart varied in an irregular manner on various timescales, but there was no correlation with the X-rays. When further optical work found Scorpius X-1 to be a member of a binary system, the initial supposition was that the primary star was transferring gas to a white dwarf and the X-ray emission came from this material.

Another scenario became available following the discovery in 1967 of the first 'pulsar' by radio astronomers Jocelyn Bell and Anthony Hewish in Cambridge. It proved to be a rapidly spinning neutron star representing the collapsed core of a supernova. Scorpius X-1 is now understood to be a neutron star that is just above Chandrasekhar's limit orbiting a companion whose mass is substantially less than the Sun.

In order to survey the entire X-ray sky, it was necessary to put instruments on a satellite. To do this, on 12 December 1970 NASA launched Explorer 42, the first of the three-mission Small Astronomy Satellite series. Since the launch was from a platform off the coast of Kenya when that nation was celebrating the anniversary of its independence, the spacecraft was named Uhuru, which means 'freedom' in Swahili.

Uhuru had a pair of X-ray telescopes installed back-to-back and they scanned the sky as the satellite slowly rotated. Whenever a source of X-rays entered the field of view, the detector generated an electrical signal. This was transmitted to the ground. By knowing the orientation of the satellite in space, astronomers were able to calculate the position of the source. Repeated surveys enabled the satellite to detect variations in the sources. It detected several hundred objects. Many were clearly outside of our galaxy, but a large proportion were so close to the plane of the Milky Way that they simply had to be 'local'.

An X-ray detector launched in 1964 from White Sands on a ballistic rocket had discovered a source in the constellation of Cygnus. Listed as Cygnus X-1, it was one of the strongest sources then known. Data between 1966 and 1968 established it to have a very small angular diameter on the sky and to be of variable intensity. Uhuru showed it to vary on a timescale of a fraction of a second. Albert Einstein's work indicated that nothing can travel faster than the speed of light, therefore an object can't flicker faster than the time it would take light to span it. For example, an object that is 1 light second across can't change its brightness faster than once per second. As the speed of light is 300,000km/sec, this meant the source of the X-rays in Cygnus X-1 could be no more than 100,000km across, which is only about one-tenth the diameter of the Sun.

In early 1971 the X-rays from Cygnus X-1

diminished in intensity and a radio source rose from below the threshold of detection. The 1 second of arc accuracy of the radio position meant the X-ray source was coincident with the 9th magnitude star listed in the Henry Draper catalogue as HDE 226868. A blue supergiant such as this cannot emit appreciable X-rays. Astronomers suspected that a companion was heating gas to a temperature of millions of degrees and causing it to emit the X-rays.

In 1971 Riccardo Giacconi predicted Cygnus X-1 was a black hole. In reporting this, the *New York Times* not only capitalised the novel term, it also used inverted commas in the expectation that it would be new to members of the general public.

Confirmation that HDE 226868 was associated with the X-ray source came in late 1972 from the Copernicus satellite, launched that year as the third Orbiting Astronomical Observatory. The 'spiral scan' of one of its instruments was able to localise a discrete source to within about 10 seconds of arc.

Observing independently, Louise Webster and Paul G. Murdin in England and Charles Thomas Bolton in Canada reported in 1972 that Doppler measurements of HDE 226868 implied the presence of a companion. There were no spectral lines from the secondary component, only those from the primary, which oscillated in wavelength with a period of 5.59983 days as the two components of the binary revolved around the common centre of mass.

By presuming that we are viewing the binary system 'edge on' (an inclination of 90°) and assuming a mass for the supergiant star, it is possible to calculate from the radial velocity of the visible star the minimum mass of the unseen object; its actual mass will depend on the inclination. The radial velocity curve establishes a well-defined 'mass function', a mathematical expression that involves the masses of the two stars and the inclination of the orbit. The plane is sufficiently angled to preclude eclipses by either component. For all sensible values for the mass of the supergiant and all reasonable inclinations (27–65°), the mass function required the unseen companion to exceed the maximum possible mass for a neutron star by a factor of about three. It could only be a black hole.

The first High Energy Astronomy

**BELOW** The radial velocity curve by Paul G. Murdin and Louise Webster for the star HDE 226868. The 5.6-day period identified it as Cygnus X-1.
*(Data from Nature, vol.235, pp37-38, 1972/ graphic W. D. Woods)*

HEAO 1

HEAO 2

HEAO 3

Observatory, launched in 1977, revealed the X-ray emission from Cygnus X-1 to be 'flickering' on a timescale of milliseconds, meaning the source could not exceed 300km in diameter. Only a black hole could compress 10 solar masses into a volume with that diameter.

A study of the hydrogen lines in the spectrum of HDE 226868 found the star to be surrounded by a gaseous envelope which is moving out at about 1,500km/sec. This 'stellar wind' appears to be transporting away material. Some of this outflow will pass within the gravitational influence of the black hole and be drawn into orbit around it. In fact, this scenario had been envisaged in 1964 by Salpeter when he realised the angular momentum of gas that is attracted by the gravitation of a singularity would cause it to spiral inward and form an accretion disc around the black hole.

Although the overall gravitational attraction of a black hole will be no greater than that of a star with the same mass at any given distance, in the case of a black hole the material can approach much closer to the centre, with the result that the gravitational force in the immediate vicinity of the hole can be enormous. By a well-known law, the inner part of an accretion disc will be rotating more rapidly than the periphery. Uhuru had detected X-rays from the intense frictional heating of the material that

**ABOVE** The three High Energy Astronomy Observatory (HEAO) satellites. The first, launched in 1977, mapped X-ray sources as a follow-up to the Uhuru all-sky survey. The second, launched in 1978 and named the Einstein Observatory, was capable of providing images of X-ray sources. The third, launched in 1979, made a gamma-ray map of the sky. *(NASA/GSFC)*

was spiralling through the inner portion of the accretion disc, heading for the black hole.

In December 1974, physicists Stephen William Hawking of the University of Cambridge and Kip Stephen Thorne of the California Institute of Technology composed a wager

**BELOW** Stephen W. Hawking and Kip S. Thorne. *(Hawking: NASA/GSFC; Thorne: Jon Rou)*

ABOVE **Martin Rees and Donald Lynden-Bell.** *(University of Cambridge)*

# Quasars and radio jets

In 1974, Martin John Rees and Donald Lynden-Bell in Cambridge said quasars were accretion discs surrounding supermassive black holes located in the cores of galaxies. Accretion disc processes can generate energy much more efficiently than nuclear fusion reactions. Nevertheless, quasars would need to convert between 0.02 and 20 solar masses per year to match the observed span of luminosities, and depending on the rate, with the more voracious ones being most luminous, they would easily gobble up 100,000 to 100,000,000 solar masses in 10 billion years. In the most luminous cases, the active region had to be gravitationally bound by at least 100,000,000 solar masses to preclude the pressure of the radiation from fully dispersing the accretion disc.

Roger David Blandford then noted that as plasma spirals in towards the black hole it carries with it a magnetic field. This 'connects' to the black hole, 'winding up' the field lines very tightly until they momentarily 'snap' and reconnect. While the field is 'open', plasma will

written on a single handwritten sheet. Thorne asserted that the Cygnus X-1 system held a black hole and Hawking insisted it didn't. At stake was a year's subscription for the American racy *Penthouse* to Thorne, or four years' of the British satirical *Private Eye* to Hawking. In June 1990 Hawking finally conceded and pornography started to arrive in Thorne's mail, much to the astonishment of his wife!

RIGHT **Arp 220 imaged by the Hubble Space Telescope, with bipolar jets from its central supermassive black hole added artistically using data from the Herschel Space Observatory of the European Space Agency.** *(NASA/JPL-Caltech)*

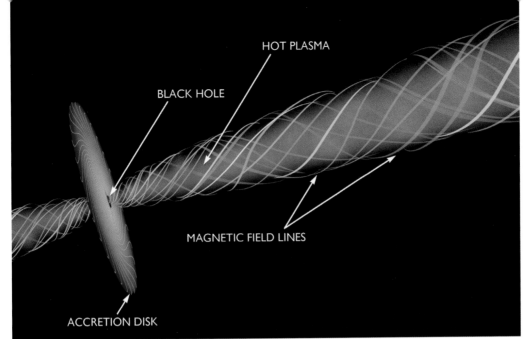

**LEFT** An explanation of how the magnetic field of a black hole forms and constrains jets of plasma. *(NASA/ESA/STScI/A. Feild/enhancement W. D. Woods)*

travel along the magnetic axes, co-aligned with the rotational axis of the black hole. Blandford established that this rapidly repeating cycle would 'pump' pulses of plasma into a 'magnetic tube', forming a jet that can be sustained for millions of years. The electrons in the plasma can be detected by radio because they are accelerated along the tube to a substantial fraction of the speed of light. Being 'hot' in the relativistic sense, the electrons emit non-thermal radiation. The magnetic wall of the tube prevents the plasma from dispersing. A jet travels out until it reaches a dense patch of the intergalactic medium, where the resulting shock wave heats that gas, dispersing it to form a radio lobe, with a 'hot spot' marking the point of contact.

If our line of sight views the galaxy more or less face-on, then we can observe the luminous accretion disc as a quasar outshining its host. However, if we view the galaxy edge-on then the accretion disc is obscured and we see the radio lobes projecting out to either side.

## Black holes in galactic cores

As the total mass of a galaxy must far exceed that of any central supermassive black hole, the presence of such an object will make an insignificant contribution to the gravity which controls the motion of the material in the outer regions but it will dominate the nuclear region. In the absence of a supermassive black hole, the stars in the nucleus would swarm about like bees. But if a black hole is present, it will draw material in and angular momentum will cause that to create a rapidly rotating disc, just as in the case of a stellar-mass black hole, but far larger. A disc of either gas or stars at the dynamical centre of a galaxy is therefore a tell-tale sign of a supermassive black hole, because without that intense gravitational field the disc would disperse.

In principle, the rotation profile of the material occupying the nuclear region of a galaxy can be measured by orienting the slit of a spectrograph in such as way as to determine the distribution of radial velocities, but few ground-based telescopes have the resolution to isolate the nuclear region of even the nearest spiral, M31 in Andromeda, which spans 1 second of arc. Nevertheless, Alan Michael Dressler at the California Institute of Technology decided in 1983 to search for supermassive black holes.

In 1943, Carl Keenan Seyfert released a catalogue of a dozen otherwise normal spirals displaying unusually bright nuclei. The most prominent of these was M77, also designated NGC 1068 and associated with the Cetus A radio source. Others were found later, some being radio sources. One, 3C120 in Camelopardalis, bore a striking resemblance to a quasar but the very small redshift prompted speculation that Seyferts (as they became known) might be the 'missing link' between quasars and normal galaxies. If one were to be viewed from afar then (it was argued) only the nucleus would be visible and it would be classified as a quasar. Seyferts

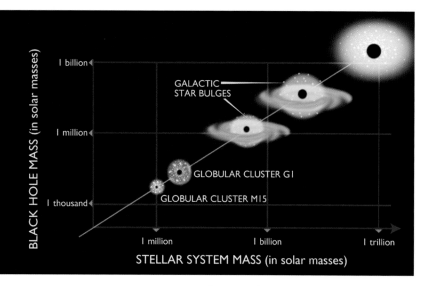

BLACK HOLE MASS (in solar masses)

I billion

I million

I thousand

GALACTIC
STAR BULGES

GLOBULAR CLUSTER GI

GLOBULAR CLUSTER MI5

I million          I billion          I trillion

STELLAR SYSTEM MASS (in solar masses)

**ABOVE There is a linear relationship between the mass of the black hole at the core of a globular cluster or a galaxy and the overall mass of its host.** *(Data from NASA/ESA/STScI/A. Feild/annotation W. D. Woods)*

were subsequently redefined as a subtype of a broader group of 'active nuclei' galaxies, most of which were ellipticals possessing jets.

M77 was a strong candidate for having a central black hole, so Dressler made it his first target. For comparison purposes he decided to inspect M31, which he had no reason to think possessed one. As events transpired, M77 was too far away for the 200in telescope to probe its nuclear region, but Dressler was astonished to find a disc of stars deep in M31 circulating at 150km/sec. This implied the presence of a hole of about 30 million solar masses. The existence of a supermassive black hole in the heart of a galaxy which showed no sign of ever having been 'active' was a major revelation. Follow-up studies by other astronomers found black holes of up to several billion solar masses in other galaxies.

In 1994, after astronauts had installed a corrective optics package to remedy the spherical aberration that marred the main mirror of the Hubble Space Telescope, it was able to resolve the cores of more distant galaxies and identify the presence of supermassive black holes. The rate of discoveries increased after the telescope was fitted with an imaging spectrograph in 1997, and black holes were soon being identified in all sorts of galaxies.

The various morphologies of galaxies derive from dynamical considerations. A galaxy such as our own, or M31, has an almost spherical nuclear region (named a 'bulge') whose centre coincides with that of a broad disc of gas and dust and the stars that coalesced from it. An

elliptical galaxy has no disc, it is 'all bulge'. This is not to say that ellipticals are smaller than spirals; giant ellipticals such as M87 are considerably larger than most spirals.

As more results were obtained, it was found that the larger galaxies had more massive central black holes. In fact, there was a linear relationship between the mass of the black hole and the luminosity (and hence the mass) of the bulge of the host.

Interestingly, a relationship of this sort had just been predicted by Joseph Silk and Martin Rees. The conventional view was that the primordial gas produced in the Big Bang had condensed to produce stars and galaxies, and the point at issue had been whether this process had run in a top-down (galaxies first) or bottom-up (stars first) manner. However, Silk and Rees proposed that the centre of each primordial gas cloud collapsed to form a black hole which, as it fed on the nearby gas, 'switched on' as a quasar. The radiation pressure then induced intense shock waves which in turn stimulated prodigious star formation. Galaxy formation was a direct consequence of the presence of the black hole. The quasar would 'switch off' when the black hole 'starved', which Silk and Rees argued would occur when the quasar was so luminous that the pressure of its radiation drove the rest of the surrounding material beyond the black hole's gravitational reach.

## Our own galaxy

Optical astronomers are unable to see the central region of our own galaxy due to the intervening gas and dust. The first insight into its nature was the realisation in the late 1930s by Grote Reber and Otto Struve that it was directly observable at radio wavelengths. Since the galactic centre lies in the constellation of Sagittarius, the strongest radio source in that direction was labelled Sagittarius A. For many years it remained the province of radio astronomers.

In 1974 Bruce Balick and Robert Lamme Brown used radio telescopes at Green Bank in West Virginia as an interferometer with sufficient resolution to identify a strong point-like synchrotron source in the heart of Sagittarius A.

They named it Sagittarius A* (abbreviated to Sgr A* with the pronunciation 'A-star').

The galactic centre appeared on the all-sky X-ray survey by the Uhuru satellite, but it was not until the launch of HEAO-2 in 1978 with an X-ray imaging system that the source was localised to within 1 minute of arc of the galactic centre. The variability of the X-rays implied a source approximately 3 light years across.

Meanwhile, a gamma-ray detector on a high-altitude balloon found a source in the general direction of the galactic centre with an energy spectrum that indicated the 'annihilation' of electrons by antielectrons. A balloon could take data only for a short period. Sustained observations by the HEAO-3 satellite, launched in 1979, found the source to vary on a time-scale that meant its diameter could not exceed several light months. The gamma-ray data was consistent with the presence of a supermassive black hole, but was not proof of it.

The Very Large Array in Socorro, New Mexico, first turned its array of 27 radio dishes towards Sagittarius A in 1981. The high angular resolution of the telescope revealed the presence of a spiral of hot gas centred on Sgr A*. After the position of Sgr A* had been refined, its position was repeatedly measured in order to place a limit on its mass, the rationale being that if it was comparable to a star it would

be seen to travel rapidly in an orbit around the centre of the galaxy. After 16 years of observing Sgr A* to be motionless, it was established to be right at the dynamical centre of the Milky Way system.

In making pioneering observations in the near-infrared in 1968 Gerhart 'Gerry' Neugebauer had discovered a star cluster listed as IRS 16 that was later found to be coincident with Sgr A*. In 1986 an improved detector resolved the cluster into red giants packed together at the equivalent of a million stars per cubic light year.

The most direct way to establish whether Sgr A* was a black hole was to study the motions of nearby streamers of ionised gas. While the radial velocities implied that 6,000,000 solar masses were concentrated within 10 seconds of arc of the Sgr A* radio source, the case was inconclusive because at a distance of 25,000 light years this was quite a large volume and the gravitating mass could be the cluster which was within several light years of the centre. The only way to be certain was to measure the motions of the stars of IRS 16 using very-high-resolution infrared astrometry.

In the early 1990s, two teams began to monitor the proper motions of the stars deep within IRS 16. As the data accumulated, one particular star was seen tracing out a large portion of an elliptical orbit with a period of 15 years. Analysis of the periapsis passage in 2002 indicated a gravitating mass of 3,700,000 solar masses. In such a confined space, it could only be a black hole. The properties of a black hole are its mass, rate of rotation, and electrical charge. The radius of an event horizon depends upon the mass of the black hole. The radius of the event horizon of the black hole at the centre of our galaxy is only 5% of an astronomical unit, a measure based on the radius of Earth's orbit around the Sun.

The Sgr A* radio source is ionised gas in the inner part of the accretion disc. Although the black hole long ago consumed all the material that was within easy reach, the Chandra X-ray Observatory has seen occasional flickering that implies ongoing irregular infall of material on to the accretion disc. Fortunately for life in the galaxy, the intense radiation associated with quasar activity is unlikely ever to resume.

**BELOW** An infrared image featuring the IRS 16 star cluster which surrounds the Sgr A* radio source. *(ESO/MPR/S. Gillessen et al)* Analysis of the proper motions of these stars revealed S2 to be in a 15-year eccentric orbit that required there to be a black hole of 3,700,000 solar masses at the dynamical centre of our galaxy. The plot axes are measured in seconds of arc, based on Sgr A*. The star's periapsis passage in 2003 was at a range of 17 light hours. *(Adapted from 'A Star in a 15.2-year Orbit Around the Supermassive Black Hole of the Milky Way' by Rainer Schödel et al., in* Nature, *17 October 2002)*

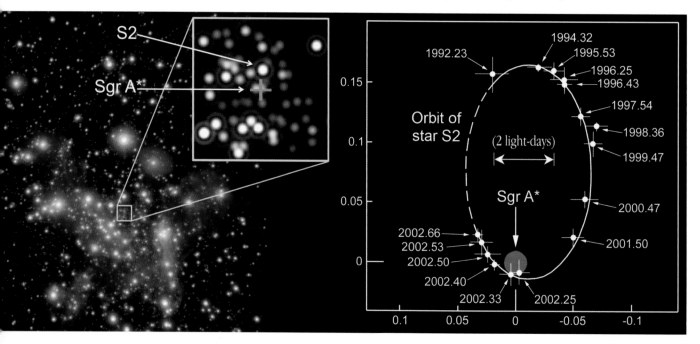

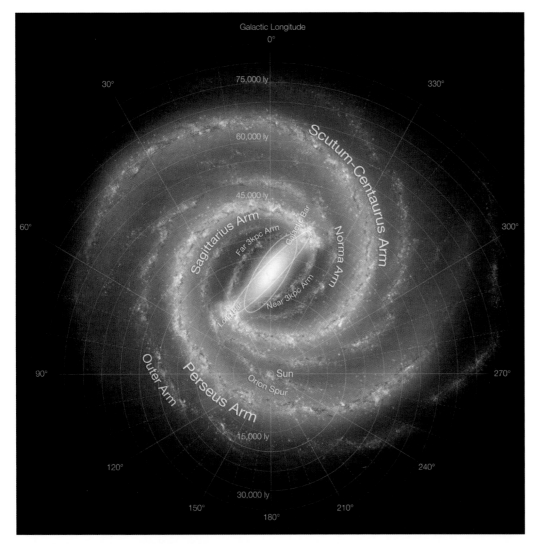

Galactic Longitude
0°

30°                                                                330°

75,000 ly

60,000 ly                                                   Scutum-Centaurus Arm

60°                                                          300°

45,000 ly

Sagittarius Arm              Far 3kpc Arm      Galactic Bar      Norma Arm

Long Bar
Near 3kpc Arm

90°                                                          270°
Outer Arm    Perseus Arm           Sun
                            Orion Spur

15,000 ly

120°                                                         240°

30,000 ly

150°                                                  210°
180°

## Chapter Six

# The age of the universe

As astronomers studied the motions of galaxies in clusters and how individual galaxies rotate, they inferred from gravitational analysis the existence of 'dark matter'. And the discovery that the expansion of the universe is accelerating indicated that a mysterious force is overwhelming the tendency of gravity to decelerate the expansion. The evidence indicates that the Big Bang occurred almost 14 billion years ago.

**OPPOSITE** When the Hubble Space Telescope took this 'ultra-deep field' in the constellation Fornax, a tiny patch of sky that appeared to be empty, it found a multitude of galaxies that dated back in time to within a few hundred million years of the Big Bang. The image, released in 2014, is a composite of exposures taken between 2003 and 2012 with the Advanced Camera for Surveys and the Third Wide Field Camera and combines visible and near-infrared light. *(NASA/ESA/H. Teplitz and M. Rafelski at IPAC-Caltech, A. Koekemoer and Z. Levay at STScI, and R. Windhorst at ASU)*

# Correcting distances

In the Second World War, Edwin Hubble was an 'army scientist'. Returning to Mount Wilson with the redshift-distance relationship extended as far as the light grasp of the 100in would permit, he had to bide his time during the development of the 200in telescope. The intention had been to commission it in the early 1940s but the work got off to a slow start and was then interrupted by the war. In 1948, even before engineers had completed refining the mirror, Hubble took a test plate and was delighted to confirm that a 10min exposure had captured faint galaxies that the 100in had been only barely able to record with an all-night exposure.

In 1948 the California Institute of Technology in Pasadena started a graduate astronomy school. One of the first students was Allan Rex Sandage, with a degree in physics from the University of Illinois. Acknowledging that he could no longer sustain a vigorous observing campaign after a heart attack in 1950, Hubble began to send Sandage to the 200in to expose plates. On Hubble's death in 1953 the task of extending the redshift-distance relationship fell to Sandage.

But first, Sandage set out to verify the distances inferred by Hubble. Using the 100in, Hubble had been able to detect Cepheid variables in only nearby spirals, so he had assumed the brightest stars in any spiral were of comparable luminosity and estimated distances on that basis. Using the 200in, Sandage found that many of the objects that Hubble had presumed to be stars were really brightly glowing clouds of hydrogen, far more luminous than a mere star.

## The inconstant Hubble Constant

The slope of the redshift-distance relationship, which became known as the Hubble Constant, was hotly debated. As Hubble had underestimated distances, it had started out rather high (in his 1929 paper reporting the relationship he gave a value of 160km/sec per million light years) and diminished as the 'distance scale' was 'corrected'.

**ABOVE** Edwin Hubble in the prime focus cage of the 200in Hale telescope. *(PalomarSkies)*

**RIGHT** **Walter Baade.** *(Mount Wilson Observatory)*

Walter Baade gained his doctorate in 1919 from the University of Göttingen in Germany, spent the 1920s at the Bergedorf Observatory in Hamburg, then joined the staff of the Mount Wilson Observatory in 1931.

As Hubble and Milton Humason examined ever more remote spirals, Baade suspected there was a flaw in the distance scale. He was particularly alarmed that our Milky Way system seemed to be far larger than the others. Also, although the globular clusters of M31 were spherically distributed around its nucleus, just as Harlow Shapley had found for those belonging to our own galaxy, they appeared to be less luminous. Another puzzle was that there did not seem to be any 'cluster variables'. Baade therefore decided to investigate M31 in detail.

Although Hubble had been able to resolve stars in the periphery of the spiral arms of M31 using the 100in telescope, his plates had shown the nucleus only as a glowing mass.

When America entered the war in 1941, Baade, still a German citizen, was an 'enemy alien'. His movements were constrained, but he was allowed to continue his astronomical work. With most of his colleagues reassigned to tasks

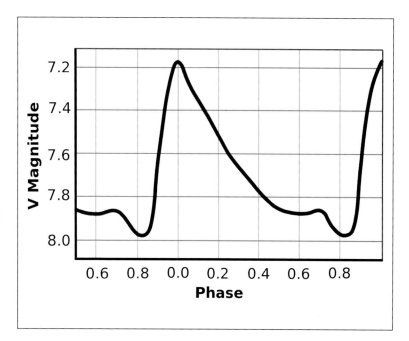

elsewhere, Baade had the largest telescope in the world more or less to himself.

In the Los Angeles 'blackout' the sky at Mount Wilson was very dark, and so Baade attempted to probe M31, still without success. Like Hubble, he had used a blue-sensitive emulsion because it was 'faster'. However,

ABOVE The light curve of RR Lyrae is typical of the 'cluster variables'.
(Wikipedia/R.J. Hall)

LEFT The Bergedorf Observatory, Hamburg, circa 1940.

molecules of air scatter most strongly at the blue end of the spectrum (which is why diffused sunlight is blue) and there was a faint glow in the night sky even in the blackout. In an effort to cut this interference, Baade switched to a red-sensitive emulsion even though it was 'slower' and therefore required longer exposures. This time his long nights at the telescope were rewarded with plates that resolved stars in the inner portion of M31. As he perfected his technique, Baade was finally able to push the telescope to its theoretical limit.

Baade found that whereas the peripheral regions of M31 were predominantly hot blue stars, the brightest stars in the nucleus were reddish. Furthermore, while the spiral arms were laden with gas and dust, the nucleus was relatively clear. It was a case of there being two 'populations' of stars. In terms of the evolutionary sequence represented by the Hertzsprung-Russell diagram, the spherical nucleus and the surrounding globular clusters were primarily older stars in the red giant or supergiant stages of their evolution. The blue stars present in the spiral arms of the flattened disc were younger. In effect, because hot blue supergiants are short-lived, the blue stars that formed in the inner regions had long

## A REMARKABLE COINCIDENCE

Walter Baade's discovery that 'classical' Cepheids were about 1.5 magnitudes brighter than their metal-poor W Virginis cousins didn't affect Harlow Shapley's 1917 method for estimating the size of the Milky Way system.

Shapley had calibrated the period-luminosity relationship found by Henrietta Swan Leavitt by calculating the mean statistical parallax of 11 local Cepheids, but several factors led him to underestimate their brightness by 1.5 magnitudes; in particular, he was unable to allow for attenuation of light by the interstellar medium because its existence wasn't announced until 1930.

The distances that Shapley calculated for globular clusters were derived from the short-period RR Lyrae variables. He had calibrated these by cluster Cepheids that were later realised to be of the W Virginis type. (In contrast, the distances to galaxies, including the Magellanic Clouds, were based on the 'classical' Cepheids that they contained.)

When Baade announced an increase of 1.5 magnitudes in the brightness of 'classical' Cepheids, this precisely cancelled out the underestimate in Shapley's calibrations!

## ASTRONOMICAL METALS

In 1929, Henry Norris Russell at Princeton investigated the surface of the Sun spectroscopically and determined that around 90% of the atoms were hydrogen, most of the rest were helium, and other elements together accounted for less than 1% of the population. Astronomers consider all elements heavier than helium to be 'metals'.

since expired. And because the gas and dust from which stars form had been consumed there was no star formation in the nucleus. In the dusty spiral arms, however, the process was ongoing. The difference in ages of the populations was evident from the fact that the young hot stars showed more 'metals' in their spectra, indicating they formed out of clouds that had been 'enriched' by the supernovae of earlier generations of stars. It transpired that these two populations of stars are also present in the case of the Milky Way system.

In 1949 Baade's first plate of M31 through the new 200in telescope confirmed his suspicion that Hubble's distance scale was flawed. If the estimated distance of just under 1 million light years was correct then he ought to have been able to detect the short-period variables of the RR Lyrae type, but they were absent at the expected magnitude. This meant either that these stars were less luminous in M31, which seemed implausible, or that it was farther away than believed.

Further work with the 200in not only identified several hundred Cepheids in M31 but also revealed that they occurred in both populations. Furthermore, their spectra were subtly different owing to a difference in composition. The 'classical' ones such as Delta Cephei in the spiral arms were 1.5 magnitudes more luminous than their metal-poor counterparts, nowadays known as the W Virginis variables, in the other population.

Hubble had used the Cepheids in the spiral arms in estimating the distance to M31 at almost a million light years, and Baade 'corrected' the distance to just over 2 million light years.

In 1952 Baade was able to announce at the

International Astronomical Union meeting in Rome that he had eliminated the issue of the anomalous luminosity of the globular clusters of M31. The RR Lyrae variables had not been evident at the predicted magnitude because the galaxy was farther away than Hubble estimated and therefore longer exposures were required in order to detect them.

By the time Milton Humason retired in 1957 he had obtained redshifts for 850 galaxies with radial velocities of up to 100,000km/sec. This sample ranged out for several billion light years and included 18 rich clusters of galaxies. In each case, a giant elliptical (often in the middle) far outshone its companions. When Sandage plotted the apparent magnitude against redshift for these 'first ranked' ellipticals they gave a straight line, implying they were equally luminous and thus could be used as 'standard candles' to probe even further. Having recalibrated the distance scale, in 1961 Sandage reduced the Hubble Constant from 55 down to 50km/sec per million light years and suspected it might be much less.

Cepheid variables were reliable distance indicators but the period-luminosity law discovered by Henrietta Swan Leavitt meant the most luminous ones had the longest periods, and it typically required at least 30 plates over a ten-year interval to verify the light curve of such a star and measure its period. When Sandage met Gustav Andreas Tammann of the University of Basel in Switzerland in 1962 they agreed to cooperate in seeking Cepheids in spirals. Tammann joined Sandage in Pasadena in early 1963 and began to analyse a backlog of plates. A year and a half later, Tammann had a dozen candidates in NGC 2403, a spiral in Camelopardalis, but their light curves were sketchy because even using the 200in they were visible only close to their peaks. Nevertheless, by 1967 the evidence supported Sandage's suspicion that the value of the Hubble Constant was still overestimated. By 1975, Sandage was beginning to think it might be as little as 15km/sec per million light years.

At this point, Gerard de Vaucouleurs of the University of Texas said Sandage was underestimating the value of the Hubble Constant by as much as a factor of two.

This kicked off a vigorous debate over 'standard candles' and their calibration. The issue was exacerbated by the fact that although Sandage's latest value was just 10% of Hubble's original estimate (in corrected terms), it was quoted with the same estimated error of plus or minus 15%. Worse, there was no overlap between the ranges of the values cited by Sandage and those by de Vaucouleurs. What was required was a significant improvement in the distances for the first steps on the distance ladder.

By then, the distances to 10,000 stars had been estimated by the measurement of their parallax and various indirect methods. It was not possible to significantly increase this sample using terrestrial telescopes, it could only be done from space. This was the primary objective of the Hipparcos satellite that was launched by the European Space Agency in 1989.

As Earth travelled around the Sun, it provided a baseline of 300 million km to measure parallaxes in order to triangulate the distances to nearby stars. By being above the atmosphere, even a small telescope on the satellite was able to attain an accuracy 100 times better than was possible from the ground. The results released in 1997 allowed the period-luminosity relationship for Cepheids to be recalibrated with an uncertainty in the distance of 5%. This meant the nearby Cepheids, which had been used as the basis of the scale for measuring intergalactic distances, were farther away than supposed, which in turn reduced the Hubble Constant by 10%. While this 'correction' was welcome, in itself it did nothing to reconcile the rival camps.

Determining the value of the Hubble Constant was one of the Key Projects of the Hubble Space Telescope. Soon after it was launched in 1990 the optics of the telescope were found to suffer spherical aberration, but once this flaw had been corrected by a Space Shuttle servicing mission in December 1993, long exposures provided exceedingly sharp images of the faint outer regions of galaxies. The goal of the team led by Wendy Laurel Freedman of the Observatories of the Carnegie Institution was to identify Cepheids in the Virgo cluster, the nearest rich cluster. Over time, the telescope identified some 800 such stars in 18 galaxies, spanning the period-luminosity

ABOVE Artwork of the Hipparcos spacecraft. *(ESA)*

BELOW An explanation of stellar parallax arising from the motion of Earth around the Sun, and the definition of 'parsec' as a unit of distance. *(W. D. Woods)*

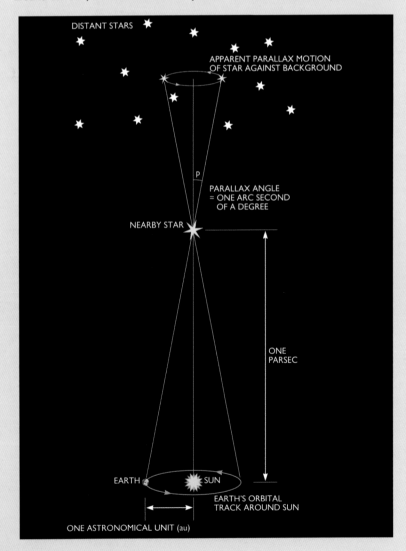

DISTANT STARS

APPARENT PARALLAX MOTION
OF STAR AGAINST BACKGROUND

P

PARALLAX ANGLE
= ONE ARC SECOND
OF A DEGREE

NEARBY STAR

ONE
PARSEC

EARTH    SUN

EARTH'S ORBITAL
TRACK AROUND SUN

ONE ASTRONOMICAL UNIT (au)

## THE HIPPARCOS MISSION

The High Precision Parallax Collecting Satellite (Hipparcos) was launched by the European Space Agency in August 1989 to undertake an astrometric survey to 12th magnitude that would measure the positions, proper motions, distances and luminosities for hundreds of thousands of stars.

The spacecraft was designed to operate in geostationary orbit but the failure of its apogee boost motor left it stranded in a highly elliptical transfer orbit. It was able to function in this orbit and provide data through to February 1993, but the software to process the data required significant revision.

It determined distances by measuring the parallax of stars across a baseline of Earth's orbit around the Sun and applying simple trigonometry. By definition, the radius of Earth's orbit of about 150 million km is an astronomical unit (au). The distance at which the angular separation of Earth and Sun (i.e. 1au) is 1 second of arc is 3.26 light years and is called a 'parsec'. The baseline for these observations of 2au therefore yielded an 'annual parallax' for each star. When integrated with radial velocities from spectroscopy, the proper motions, parallaxes, distances and tangential velocities measured by Hipparcos allowed the space motion of stars to be calculated.

Prior to the Hipparcos mission, parallax distances were known for 8,000 stars. The catalogue published in 1997 took that number to 118,218 stars with parallax measurements accurate to 1 millisecond of arc. This was sufficient to determine the distances out as far as 1,600 light years for 20,000 stars to an accuracy of 10%.

Although Hipparcos greatly enhanced our understanding of the structure and dynamics of the Milky Way system, its survey was very localised and it included less than one-millionth of the stars in the system. This was because the farthest measurable distance of 1,600 light years was comparable to the half-thickness of the galactic disc at the Sun's position, some 27,000 light years from the centre. As a result, the survey was of a small spherical volume in the 'local' part of the spiral structure.

relationship. By accurately establishing the distance to the Virgo cluster, this study calibrated the 'standard candles' that were being used to reach much farther. The value of the Hubble Constant was then calculated by a variety of methods. The 'best' estimate, announced in 2001, was 21.5km/sec per million light years to an accuracy of 10%.

By then, the values estimated by the two rival teams on the basis of terrestrial observations were just outside this range, one above and the other below, so the convergence was pleasing.

## Non-luminous matter

Fritz Zwicky gained his doctorate in physics in Switzerland in 1922 and then moved to America where, in 1925, he joined the California Institute of Technology in Pasadena. He was an irascible person who readily bore grudges and therefore tended to work alone. Nevertheless, he undertook some major projects and made significant discoveries.

The Coma cluster in the direction of Coma Berenices is estimated to be about 300 million light years away. It is well away from the obscuring gas and dust that lies in the plane of the Milky Way. The cluster contains thousands of galaxies, and is split into two clumps.

The typical separation between members is 300,000 light years, so it is very densely packed. Overall, the cluster spans more than 20 million light years.

In 1933 Zwicky investigated the Coma cluster and discovered it was not in an overall state of rotation. The constituent galaxies were moving randomly around a common centre, but the gravitational attraction that was required in

**LEFT Although Fritz Zwicky was a prickly character he did some excellent work.**

order to hold the cluster together far exceeded that of the matter which was visible.

To determine the random motions within the cluster, Zwicky measured the radial velocities of 600 members relative to the mean for the cluster as a whole, which was 7,000km/sec. The spread of velocities relative to the mean provided a measure of the kinetic energy of the system. The greater that energy, the greater would be the gravitational attraction required in order to prevent the members of the cluster from escaping. Next, Zwicky estimated the mass that was available to cause gravitational attraction. For this, he used the mass-luminosity relationship in which the more luminous a galaxy, the greater the number of stars generating the light. From the light that was captured by his plates, he estimated the mass of each galaxy. On adding up, the overall mass was insufficient to account for the gravitational attraction inferred from the kinetics.

For the cluster to remain intact, there had to be 50 times more material than was visible as 'luminous mass', so he coined the term 'non-luminous' matter.

Despite its profound implications, Zwicky's discovery was largely ignored by the astronomical community, at least until another line of research prompted the same inference.

On joining the Carnegie Institution of Washington DC, Vera Rubin worked with William Kent Ford to construct an innovative spectrograph with an image-intensifier capable of capturing the spectra of faint objects. Ignoring the popular frenzy to study quasars and active galaxies, Rubin decided to investigate normal galaxies. Whereas Allan Rex Sandage integrated light from an entire galaxy in order to measure its mean redshift, this new spectrograph was used to measure the radial velocity across the diameter of a galaxy. The velocity distribution along the major axis, known as a 'rotation curve', is a measure of the orbital velocity of stars or gas clouds at different distances from the centre, and provides a profile of the distribution of mass in the galaxy.

In the solar system, where 99% of the mass is centrally located in the Sun, the planets obey Johannes Kepler's laws and their orbital velocities decrease with the square root of the radius of their orbit; a profile known as 'Keplerian decline'. A spiral galaxy has a characteristic distribution of light that fades exponentially out from the centre. It had been presumed that the distribution of luminosity would be correlated with the distribution of mass because, after all, the light was coming from stars. The rotation curve of a spiral galaxy was therefore expected to display Keplerian decline.

Rubin selected M31 as the first objective of

study simply because it subtended such a broad angle on the sky that the slit of the spectrograph could isolate small sections of it, to measure the rotation curve in fine detail. The observations, made at the Kitt Peak National Observatory in Arizona, were a shock: the rotation curve was 'flat'. After rising rapidly deep within the nucleus, it slowly attained a level that persisted far into the periphery where, even though the luminosity declined, the mass evidently did not. When she reported this surprising finding in 1970 it attracted a sceptical reaction.

Undeterred, Rubin made a systematic study of the rotation curves of different types of spiral, in every case finding the outer regions of the galaxy to be rotating as if they were contained within a larger non-luminous mass.

Support came from computer modelling of how galaxies formed and evolved. Phillip James Edwin Peebles and Jeremiah Paul Ostriker at Princeton discovered in 1973 that the spiral form, which was really nothing more than a 'density wave', was readily disrupted unless its disc resided within a larger spherical halo whose gravity maintained its structure.

Rubin's rotation curves were obtained from optical spectra, so they extended out only as far as the disc of visible stars.

In 1978 Albert Bosma at the University of Gröningen reported 21cm emissions by neutral hydrogen in 20 spirals obtained by the radio telescope at Westerbork in the Netherlands, showing that the 'flat' rotation extended much farther out than the visible disc.

The density of neutral hydrogen in a spiral declines steadily with increasing distance from the centre, but the observations showed a sharp cut-off in the 21cm emission. This did not mean there was no hydrogen farther out, just that if it were present it wasn't emitting at that wavelength and so didn't consist of cold, neutral atoms. The ultraviolet light which pervades intergalactic space could easily ionise hydrogen. In a sea of plasma, free protons and electrons will meet and recombine, then be re-ionised in an endless cycle. The result will be emission at wavelengths characteristic of recombination, notably hydrogen-alpha. Although this emission is in the optical spectrum, that originating from the peripheral regions of galaxies would be faint. Nevertheless,

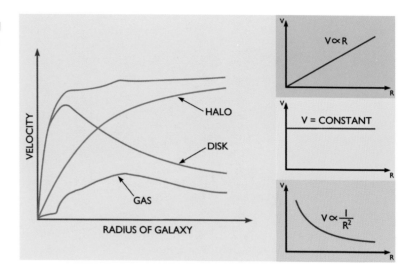

recent improvements in spectroscopy have enabled it to be detected and the results indicate that the 'flat' rotation extends beyond the cut-off of neutral hydrogen.

Spirals, as Walter Baade had highlighted, combine a spheroidal population of old, metal-poor stars which doesn't rotate as a whole, where the orbits of the stars are elliptical and at all angles, and the concentric disc in which the stars trace out essentially circular orbits. Since an elliptical galaxy comprises only the spheroidal component, there is no systematic rotation to measure as a way of calculating the gravitating mass. However, the motions of outlying globular clusters of ellipticals provide an

**ABOVE** A schematic rotation curve for a galaxy (left, based on data by Vera Rubin) and what would be expected for a solid body, for a constant velocity, and for Keplerian decline. *(W. D. Woods)*

**LEFT** Jeremiah P. Ostriker. *(Princeton)*

**ABOVE** George Abell and the 48in Schmidt survey telescope of the Palomar Observatory. He conducted a study of clusters of galaxies. *(Caltech)*

**BELOW** This Hubble Space Telescope image shows that the mass of the rich cluster of thousands of galaxies called Abell 2218 in the constellation Draco is 'lensing' the light from more remote sources, magnifying it and stretching it out into arcs. Multiple distorted images of the same galaxies can be identified by comparing their shapes and colours. In addition to the giant arcs, many smaller arclets have been identified. *(NASA/ESA/Johan Richard at Caltech and Davide de Martin and James Long at ESA)*

estimate of the gravitating mass. For example, M87, the giant elliptical at the heart of the Virgo cluster, has thousands of globulars. The results indicated that ellipticals, too, are surrounded by non-luminous mass.

Further studies suggested that 'dark matter' pervades galaxy clusters, just as Fritz Zwicky had inferred for the Coma cluster.

One prediction of Albert Einstein's theory of General Relativity was that space must be distorted by the gravitation of large masses. He was proved correct by Arthur Eddington's measurement of how the light from stars was 'bent' when it passed close to the Sun on its way to Earth during a solar eclipse in 1919.

On a larger scale, it was evident that when the light from a very remote galaxy passed close by an intervening galaxy on its way to us, it, too, would be bent. In fact, if the alignment was just right, it was even possible that the gravitation of the nearer galaxy would act as a lens and produce an image of the background object, albeit distorted. It was realised that if galaxy clusters contain a great deal of dark matter, that, too, would function as a lens.

Abell 2218 is a rich cluster of galaxies at a distance of around 3 billion light years. When it was listed for imaging by the Hubble Space Telescope, the cluster had already been revealed by ground-based astronomy to be a gravitational lens.

On seeing the result from the space telescope in September 1994, the scientists were amazed by the clarity of the view.

There were a number of thin arcs of light on the line of a circle, each of which was a distorted view of the same background galaxy, located three or four times farther away and thus too faint to be seen directly. Its presence was revealed only by the magnification of the intense gravitational field of the foreground cluster of galaxies. By precisely plotting the lensed artefacts and then working backward, it was possible to calculate the gravitating mass responsible for the lensing, and this could then be compared with what was inferred to be present by normal means.

First, the masses of the galaxies could be calculated from the amount of light which they produced. Then it was necessary to determine the amount of gas that occupies the voids between the galaxies. Just as galaxies in clusters travel around the centre of gravitation in a state of equilibrium, so does the gas that lies between them. The temperature of gas in gravitational equilibrium is directly related to the total gravitating mass. In large clusters like the Virgo cluster, with approximately a thousand-trillion solar masses, the temperature of the gas situated between the galaxies can reach 100 million degrees, making it detectable by its X-ray emission. It turns out that the total mass of such gas can be five to ten times that of the galaxies. Most of the normal matter in a cluster is therefore primordial hydrogen that was not incorporated into galaxies. The fact that the mass inferred from gravitational lensing greatly exceeds the normal matter in the cluster proves it is dominated by dark matter.

Astronomers had naturally presumed that the universe comprised what they could see, little suspecting that the luminous material was insignificant!

## The value of Omega

Knowing the value of the Hubble Constant enables us to extrapolate back to a time when all the galaxies were together, and so calculate the age of the universe. But this presumes the expansion rate has been constant, which it can't have been because the mutual gravitational attraction of the mass of the galaxies must act to slow the expansion.

Any variation in the rate of expansion would reveal itself as a difference in the value of the Hubble Constant over time. Although Hubble himself had sought a departure from linearity at the far end of the redshift-distance relationship there was considerable scatter in his data, and in any case his survey did not extend far enough into space.

Even after Sandage had extended the relationship out to a recessional velocity of 250,000km/sec, in 1974 he reported no clear indication of divergence.

At this point, John Richard Gott and James Edward Gunn at the California Institute of Technology and Beatrice Muriel Tinsley and David Norman Schramm at the University of Texas decided that instead of trying to compete

with Sandage they would measure the Density Parameter, known as Omega, which is the mean mass-energy density of the universe expressed as a ratio of the critical density. A value of 1.0 would indicate a 'flat' universe. Estimating Omega involved thinking on the largest of scales.

Gott, Gunn, Tinsley and Schramm used three methods to estimate Omega. The total luminosity of galaxies enabled the mass of the stars that emitted the light to be estimated. This yielded a value for Omega of 0.01. By analysing the motions of galaxies in clusters they got a value for Omega of 0.1. Finally, they calculated the deuterium abundance from the absorption lines in interstellar space, measured by the Copernicus Orbiting Astronomical Observatory. Their rationale for this latter method was that the ratio of deuterium to helium would have been very sensitive to conditions in the primordial fireball. In effect, deuterium measured the mass-energy density in the early universe,

**ABOVE** Clockwise from top left: **John Richard Gott** *(Princeton)*, **Beatrice Muriel Tinsley** *(The Encyclopedia of New Zealand)*, **David Norman Schramm** *(University of Chicago)* **and James Edward Gunn** *(Princeton).* **They used a variety of astronomical methods to determine the value of Omega.**

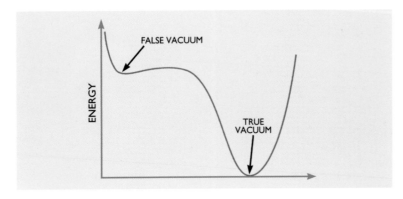

ABOVE Alan Guth realised that the cooling primordial fireball spent a brief period in a highly energised 'false vacuum', the 'negative pressure' of which induced exponential expansion. *(W. D. Woods)*

BELOW Shortly after the Big Bang, the universe underwent a brief period of exponential expansion that doubled its radius 100 times, inflating it from a volume no greater than that of a proton to about the size of a grapefruit. This smoothed out the energy fields and flattened space, making Omega precisely 1.0. After that the expansion proceeded at the rather more sedate rate indicated by the Hubble Constant. *(W. D. Woods)*

ABOVE Alan Guth.

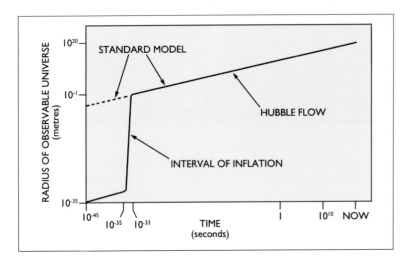

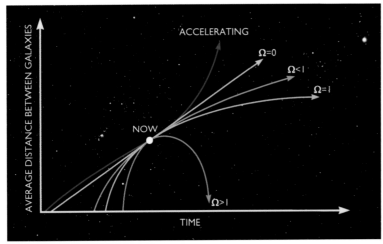

rather than as it is today. This, too, yielded a value for Omega of 0.1. Overall, therefore, their results in 1974 strongly indicated that the universe must be 'open'.

Although 90% of the gravitating mass-energy would have to be 'missing' for the universe to be 'flat', there was a compelling basis for believing this to be so. In 1979 Robert Henry Dicke and Phillip James Edwin Peebles at Princeton argued that the slightest deviation from unity in the value of Omega at the origin of the universe would have caused it to rapidly become either infinitesimally small or infinitely large. For Omega to be anywhere near the measured value, it must have been precisely 1.0 at the start.

In 1979, Alan Harvey Guth at Cornell calculated that shortly after the Big Bang there would have been a very brief period of exponential expansion as the rapidly cooling fireball lingered in a 'false vacuum', and during that interval the radius of the universe would have doubled 100 times, causing it to expand from a volume no greater than that of

LEFT The value of Omega determines the fate of the universe. *(W. D. Woods)*

a proton to a sphere with a radius of 10cm. This 'inflation' would have efficiently 'flattened out' any variations in density and made Omega precisely 1.0.

While the Hubble Space Telescope was making observations to determine the value of the Hubble Constant by observing nearby clusters of galaxies, two teams of ground-based astronomers set out to extend the redshift-distance relationship using Type Ia supernovae as 'standard candles' to probe the early universe. They were seeking divergence from linearity as a measure of how the rate of expansion of the universe had slowed down.

One team was led by Saul Perlmutter of Berkeley, California, and the other by one of his former postdocs, Brian Paul Schmidt, who was then at Siding Springs in Australia. In 1998 they independently announced a progressive departure from linearity in the relationship at very high redshifts which indicated that the rate of expansion was *accelerating*, not decelerating. This result was utterly unexpected. It meant the mutual gravitational attraction that acted to slow the expansion of the universe was being overwhelmed by a universal repulsive force of some kind.

As regards Omega, the Wilkinson Microwave Anisotropy Probe was launched by NASA in 2001 as a follow-up to the Cosmic Background Explorer.

## TYPE IA SUPERNOVA

This supernova is a runaway nuclear detonation which completely destroys a white dwarf that has attained Chandrasekhar's limit by accreting material from a companion star in a binary system.

As the situation is very specific, the prodigious release of energy is always the same and the light curve is very distinctive. A supernova of this type can be seen across billions of light years. By knowing the true luminosity, the peak apparent magnitude gives the distance.

**BELOW** Saul Perlmutter of University of California, Berkeley (left) and Brian Schmidt of Siding Springs, Australia, independently discovered that the rate at which the universe is expanding is not slowing down (as expected) but accelerating. *(Respectively: Roy Kaltschmidt at Lawrence Berkeley National Laboratory, and Tim Wetherell)*

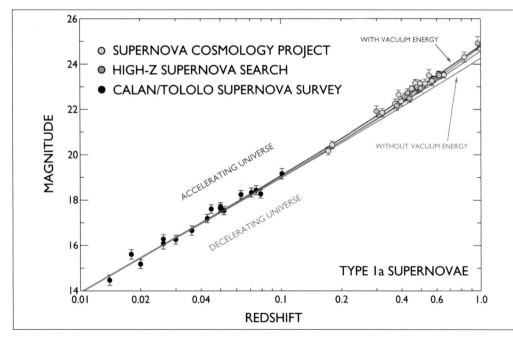

SUPERNOVA COSMOLOGY PROJECT
HIGH-Z SUPERNOVA SEARCH
CALAN/TOLOLO SUPERNOVA SURVEY

WITH VACUUM ENERGY

WITHOUT VACUUM ENERGY

ACCELERATING UNIVERSE

DECELERATING UNIVERSE

TYPE Ia SUPERNOVAE

MAGNITUDE

REDSHIFT

**LEFT** Plotting redshift versus brightness for Type Ia supernovae in distant galaxies revealed that the rate at which the universe is expanding is accelerating, rather than, as had been presumed, decelerating. *(Saul Perlmutter, 'Supernovae, Dark Energy, and the Accelerating Universe', Physics Today, April 2003/annotation W. D. Woods)*

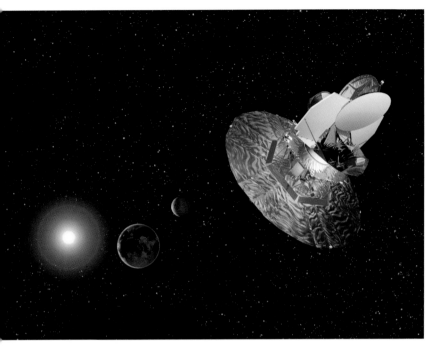

ABOVE Artwork of the WMAP spacecraft. *(NASA/WMAP Science Team)*

ABOVE RIGHT The Wilkinson Microwave Anisotropy Probe spacecraft was named after David Todd Wilkinson, a leading member of the science team, here shown shortly prior to his death in 2002. *(Princeton)*

RIGHT The full-five-year all-sky map of the cosmic microwave background produced by the WMAP mission. It represents the universe 380,000 years after the Big Bang. Anisotropies in the temperature field (red hot spots and blue cold spots) are evidence of the earliest gravitating structures. *(NASA/ WMAP Science Team)*

COBE had measured the temperature of the cosmic microwave background as 2.726(±0.01)K. But that was an average. An all-sky temperature map revealed a variation that meant there were already fluctuations in density when the radiation field decoupled. It was from such 'wrinkles' that the large-scale structures that we see today developed. The objective of WMAP was to study this earliest structure in detail. On achieving its nominal two-year mission in September 2003, the project was granted a series of extensions through to September 2010.

The initial results from WMAP in 2003 proved that Omega is 1.0 to within the unprecedented accuracy of 1%. The dataset was updated at two-yearly intervals, and the final results in 2013 indicated that the visible matter in galaxies accounts for a mere 4.6% of the density that is needed for the measured value of Omega; that the 'dark matter' whose gravity binds clusters together accounts for another 24%; and the remaining 71.4% is not present as matter, it is the 'dark energy' that is driving runaway acceleration.

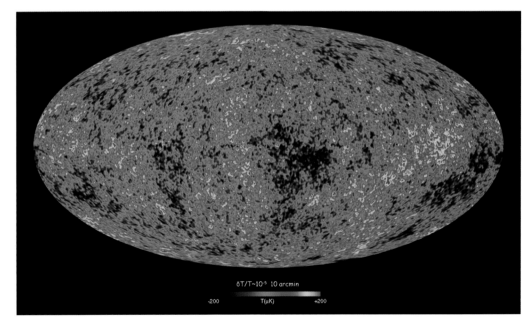

$\delta T/T \sim 10^{-5}$  10 arcmin

-200   T(μK)   +200

With the Hipparcos recalibration of the Cepheid variables, the Hubble Space Telescope's determination of the Hubble Constant as 21.5km/sec per million light years put the age of the universe at 14 billion years in the absence of deceleration. Taking into account the value of Omega as 1.0 and allowing for deceleration, but not acceleration, the age would be 10 billion years. Allowing for dark energy, the rate of expansion would initially have decreased owing to deceleration while the galaxies were tightly packed, but, as

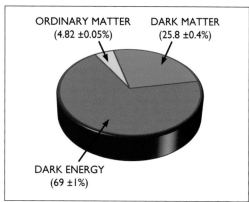

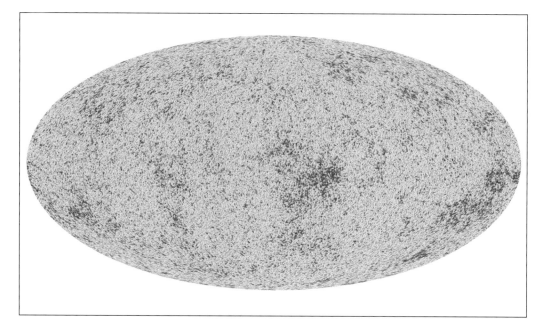

they thinned out, the repulsive force became dominant about 5 billion years ago. Taking everything into account, this gave an age for the universe of 13.7 billion years.

In 2009 the European Space Agency's Planck mission began to compile an all-sky map of the cosmic microwave background to investigate the finer detail of the WMAP data. The results in 2013 placed the age of the universe at 13.798(±0.037) billion years and gave a composition of 4.82(±0.05)% ordinary matter, 25.8(±0.4)% dark matter, and 69(±1)% dark energy.

At this point, it is necessary to explain that although astronomers use the terms 'dark matter' and 'dark energy' they do not know what these things are, although there are a variety of possible candidates. The names are mere placeholders until the day, hopefully not too far off, that we come to understand their nature better. We are aware of their existence only by their influence on the large-scale structure of the universe.

# A fluctuation of the vacuum

Edward Tryon gained his PhD at the University of California at Berkeley on the relationship between General Relativity and Quantum Field Theory. In 1973 he realised that the gravitational potential energy of all the matter in the universe is equal and opposite to its mass energy.

To explain, if two stationary objects in space are attracted by gravitation they will move closer together. In motion, they gain kinetic energy.

BELOW **The first page of Ed Tryon's paper suggesting that the universe is a random fluctuation of the vacuum.** (Nature, 14 December 1973)

This means that in their initial positions they possessed a *potential* energy. As they draw together, the potential energy is converted to kinetic energy. By convention, we say that objects have zero potential energy when their separation is infinite. For potential energy to decrease as objects are drawn together, it must become negative. And as Albert Einstein established, matter is simply concentrated positive energy. These balance out and the net energy of the universe is precisely zero.

Because Quantum Field Theory requires only that the average energy of the vacuum be zero, it is entirely plausible that the universe constitutes a 'quantum fluctuation'. In that case, not only was the universe created from nothing, it is, in essence, still nothing.

Tryon reasoned that the probabilistic nature of Quantum Field Theory, which says that anything which isn't forbidden will eventually occur, made it inevitable that the universe would be created spontaneously from the vacuum.

The idea of the universe originating as a random fluctuation of the vacuum is speculative, but the fact that its net energy is zero is profound.

# The observable universe

If a galaxy is so highly redshifted that we see it as it was early in the universe, then its light will have taken all that time to reach us. We could reasonably say we were looking toward a 'horizon'.

If we were to observe such a galaxy in one direction and another one in the opposite direction, then although the light from one of the galaxies has had time to reach us, it is only halfway to reaching the galaxy on the opposite side of the universe. A sphere whose radius is the distance that light has travelled since the beginning of time therefore defines the 'observable' universe. As we are not in a privileged position at the centre of the universe, there will be a similar sphere for every point. So an observer on one of the galaxies that we see at the edge of the universe will see us similarly positioned, and will see a part of the universe that is beyond our horizon. Note that the horizon alluded to here is not one of distance, but of time. We do not see distant galaxies

396

NATURE VOL. 246 DECEMBER 14 1973

## Is the Universe a Vacuum Fluctuation?

EDWARD P. TRYON

Department of Physics and Astronomy, Hunter College of the City University of New York, New York, New York 10021

The author proposes a big bang model in which our Universe is a fluctuation of the vacuum, in the sense of quantum field theory. The model predicts a Universe which is homogeneous, isotropic and closed, and consists equally of matter and anti-matter. All these predictions are supported by, or consistent with, present observations.

Universe consists equally of matter and anti-matter has been studied by many authors; for a recent review see ref. 3.)

Of the remaining conservation laws, the most important for cosmology is that concerning energy: although matter and energy can be converted into each other, the net energy remains constant if an intrinsic energy of $mc^2$ is assigned to each piece of matter.

The Universe has an enormous amount of mass energy, and this might be thought to preclude a creation of the cosmos from nothing. There is, however, another form of energy which is important for cosmology, namely gravitational potential energy. The gravitational energy of a mass $m$ due to its interaction with the rest of the Universe is given roughly by

$$E_g \approx -GmM/R$$

where $G$ is the gravitational constant and $M$ denotes the net mass of the Universe contained within the Hubble radius $R = c/H$, where $H$ is Hubble's constant.

### Quantum Field Theory

To indicate how such a creation might have come about, I refer to quantum field theory, in which every phenomenon that could happen in principle actually does happen occa-

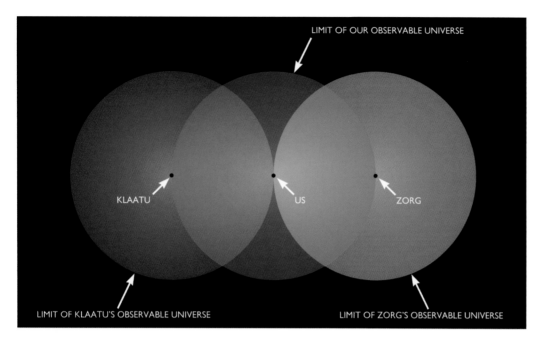

LIMIT OF OUR OBSERVABLE UNIVERSE

KLAATU

US

ZORG

LIMIT OF KLAATU'S OBSERVABLE UNIVERSE

LIMIT OF ZORG'S OBSERVABLE UNIVERSE

**LEFT** The limitation imposed by the speed of light means the 'observable universe' is a small fraction of the universe. Due to the time which light takes to reach us, the objects on our 'horizon' are seen as they were in the remote past. An observer in a galaxy which lies on our horizon will see our galaxy on their horizon as it was long ago. As a result, although we can see the galaxies in which Klaatu and Zorg live, and they both can see us, they cannot see each other.
*(W. D. Woods)*

as they are now, we see them as they were long ago. And farther back in time, beyond any galaxies, there is the cosmic microwave radiation field with an enormous redshift.

## The ultimate theory

A particle and its antiparticle can be spontaneously created out of energy, but that energy must be at least equal to the sum of the 'rest masses' of the partners in accordance with Albert Einstein's equation, $E=mc^2$. If the two particles meet, they will convert back into energy.

The Uncertainty Principle that the German theoretical physicist Werner Karl Heisenberg discovered in 1927 says that the more precisely we define the position of a particle, the less precisely we know its momentum (energy), and vice versa. On the smallest of scales, the uncertainty in the energy can become so significant that it is possible for particle/antiparticle pairs to spontaneously appear in what is otherwise an utterly empty vacuum. But these 'virtual' particles rarely make their presence felt; they simply flit in and out of existence. As Quantum Field Theory is based on the Uncertainty Principle, it requires the vacuum to be literally seething with energy.

The noted American theoretical physicist John Archibald Wheeler succinctly summarised Einstein's General Relativity by saying, "Space-

time tells matter how to move, and matter tells space-time how to curve." Consequently, in the absence of significant mass (such as in the vacuum of deep space), space must necessarily be 'flat' and smooth; the gravitational field should be zero, and there should be no energy at all.

And therein lies the contradiction between the two theories, each of which is spectacularly successful in its own realm. Whereas General Relativity requires the gravitational field on the scale of the quantum realm to be zero, the Uncertainty Principle requires only that it maintain an *average* of zero. Because gravitation is a measure of the curvature of space, if its value is fluctuating wildly, space, far from being smooth, will be so warped as to constitute a random froth which Wheeler described as "quantum foam". Consequently, the equations of General Relativity produce infinites. These arise from the fact that Quantum Field Theory deals with fields that originate from dimensionless points.

It is apparent that neither Quantum Field Theory nor General Relativity is a complete description of reality; each is actually *an approximation* in its own realm of something more fundamental that applies at all energies, including those in the first moments after the Big Bang. Despite decades of head-scratching by the finest minds on the planet, we don't yet have such a theory.

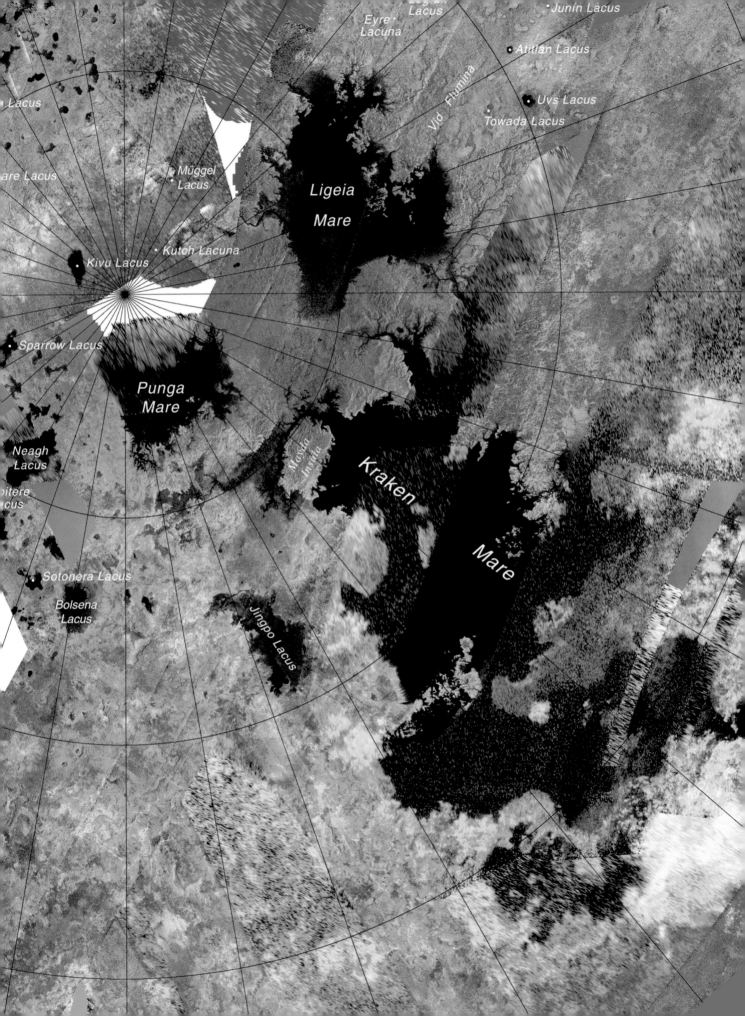

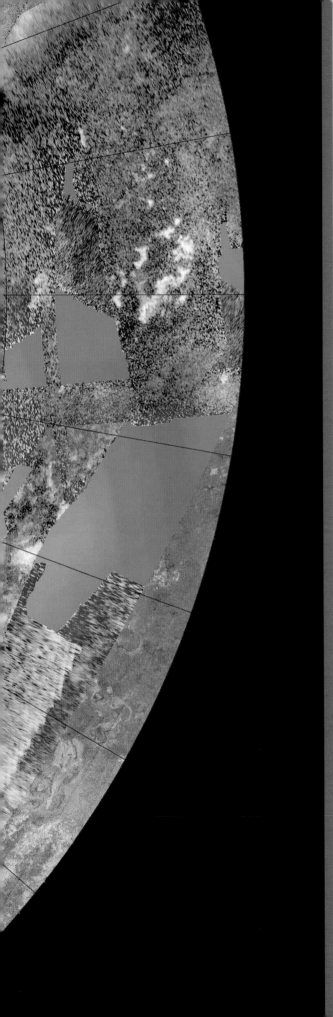

*Chapter Seven*

# The solar system

A remarkable aspect of the universe is that it provides conditions conducive to life. This chapter reviews our changing understanding of the prospects for extra-terrestrial life in the solar system. One surprising discovery from robotic space missions is that there are oceans under the icy shells of bodies in the outer solar system, and that these might host microbial life.

**OPPOSITE** Some of the lakes of hydrocarbons in the north polar region of Titan, the largest moon of Saturn, are large enough to be classified as seas. *(NASA/JPL-Caltech/ASI/USGS)*

## 127

## An outline

In general terms, in addition to the centrally located Sun, the solar system consists of four small rocky inner planets and four large giants, all of which travel in more or less circular orbits and in more or less the same plane. Many of these have a collection of moons which accompany them around the Sun. In addition, there are innumerable smaller bodies in a wide variety of orbits, some confined to broad circular 'belts' and others in elliptical paths that cross the orbits of planets.

The inner belt is located between the orbits of Mars and Jupiter. It is known simply as the 'asteroid belt'. The outer belt lies beyond

the orbit of Neptune, the outermost giant, and is named after two astronomers who independently studied it, Kenneth Essex Edgeworth and Gerard Peter Kuiper. It is usually referred to as the Kuiper belt. Most notable of the minor bodies are the 'dwarf planets' Ceres in the inner belt and Pluto in the outer belt.

The largest planet, Jupiter, is a gas giant some 318 times as massive as Earth. In fact, it is more massive than the rest of the Sun's retinue added together. Most of it seems to be an envelope of hydrogen and helium. Saturn, lying farther out, is somewhat smaller. The outer giants, Uranus and Neptune, are believed to possess cores that include a mix

of icy materials and their envelopes aren't so dominated by hydrogen. To distinguish them from the gas giants they are often referred to as ice giants.

Most of the small bodies in the inner solar system are at least partially made of rock and those in the outer system are primarily ice. In a class of their own are the comets. These are mainly icy bodies. Beyond the Kuiper belt there seems to be a broad disc of icy bodies that form a source of comets that enter the inner system close to the plane in which most other bodies travel. And beyond that, almost halfway to the nearest star, there is a spherical shell of icy bodies that can fall into the inner system at any angle.

# Ideas about origins

The French naturalist Georges-Louis Leclerc, Comte de Buffon, suggested in 1745 that the planets condensed from the material which was left after a comet collided with the Sun. He did not know that a comet, although voluminous, is a tenuous gas and hence insignificant. The prospect of a comet hitting the Sun had seemed to him to constitute a titanic event. In such a hypothesis, the development of a system of planets around a star would be a chance event.

In contrast, in 1755 the German philosopher Immanuel Kant proposed that the Sun and its retinue of planets were created when an interstellar cloud of gas underwent a process of gravitational collapse. In this case, all stars might possess planetary systems.

However, such theories had to be judged on how well their mathematics was able to explain observations. Pierre-Simon Laplace, the premier mathematician of his time, set out to investigate how a cloud of gas might collapse. He immediately rejected Buffon's idea for its lack of mathematical rigour.

In his 1796 book *Essay on the System of the World*, Laplace noted that all of the planets are in almost circular orbits, that they travel around the Sun in the same direction and in almost the same plane, and (as far as he knew) they turn on their axes in the same direction. He argued that as the cloud of gas condensed, angular momentum would have made it rotate at an accelerating rate which would have prompted it to shed 'excess' momentum by ejecting an equatorial ring of material. However, as the collapse continued, the rotation rate would have increased again. By the time the proto-Sun reached a stable state, it would have shed a number of concentric co-planar rings. These would have individually coalesced to produce a number of planets travelling in near-circular orbits. And if this process were to be repeated on a smaller scale, the planets would develop systems of moons.

In 1845 William Parsons in Ireland, armed with a telescope that had a mirror measuring an unrivalled 72in in diameter, found the M51 nebula to have a spiral structure. Unable to resolve any stars, he inferred this to be a disc of gas swirling around a nascent star and therefore proof of Laplace's 'nebular hypothesis'.

Further support came in 1854 when the German physicist Hermann Ludwig Ferdinand von Helmholtz suggested the Sun derived its tremendous energy from ongoing gravitational collapse.

On the other hand, a mathematical analysis by the Scottish physicist James Clerk Maxwell found in 1859 that the planets could not have coalesced from rings of material shed by the proto-Sun.

By the end of the 19th century, Laplace's scheme was also in trouble because of other dynamical issues. In particular, it was difficult to explain why the planets ended up with most of the system's angular momentum. The planets account for barely 0.1% of the mass of the system but possess 98% of its angular momentum. In fact, Jupiter, which is larger than all the other planets combined, has 60% of the solar system's angular momentum. Also, if the formation of moons involved this same process, why was most of the momentum in the Jovian system held by the planet, rather than by its moons? Laplace had never been able to explain

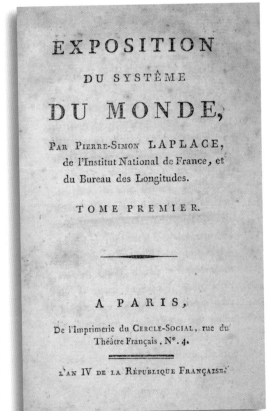

EXPOSITION

DU SYSTÈME

DU MONDE,

PAR PIERRE-SIMON LAPLACE,
de l'Institut National de France, et
du Bureau des Longitudes.

TOME PREMIER.

A PARIS,

De l'Imprimerie du CERCLE-SOCIAL, rue du
Théâtre Français, N°. 4.

L'AN IV DE LA RÉPUBLIQUE FRANÇAISE.

**RIGHT** The title page of the first edition of Pierre-Simon Laplace's *System of the World*, published in 1796.

how the angular momentum was transferred to the rings of material. Another issue that he had never been able to account for was why the Sun, which has an axial rotation of about a month, should end up rotating so slowly. Nor was it apparent why the spin axis of the Sun should be inclined at 7.25° to the mean plane of the planetary orbits.

In the 1880s Alexander William Bickerton in New Zealand suggested that the Sun had collided with another star, and the planets condensed from the material that was ejected into space. Support for this idea came with a spectacular nova in the constellation of Perseus in 1901. At that time, novae were believed to be stellar impacts. When a shell of glowing gas was later found at the position of the nova, this was taken to be proof that collisions prompted the ejection of debris.

With the turn of the new century, the renowned American geologist Thomas Chrowder Chamberlin and astronomer Forest Ray Moulton set out to investigate whether the planets had formed as a result of the Sun having a very close encounter with another star. They reasoned that an actual collision was not necessary in order to cause material to be ejected into space. At the closest point of approach the mutual gravitational attraction would first have produced a tidal bulge on each star, prior to drawing out a cigar-shaped streamer linking the stars. As the stars drew apart, this streamer would have become detached in space. As the streamer cooled, the material would have condensed to create a large number of small solid bodies which they named 'planetesimals'. These later accreted to form planets. The largest planets would have formed in the centre of the streamer and the smaller ones toward the ends.

In 1917 James Hopwood Jeans and Harold Jeffreys in England addressed the issue of how most of the angular momentum could be transferred to the planets by suggesting the other star would have drawn the gaseous stream much farther into space than would have occurred if material had simply been 'spun off' by the Sun during contraction. To overcome problems with the dynamics of the manner in which the planets would have formed, in 1929 Jeffreys concluded that the Sun must have had a grazing encounter with the interloping star.

The American physicist Lyman Strong Spitzer showed in 1939 that material drawn from the Sun would have been so hot that it would have rapidly expanded and dispersed, rather than cooling and condensing. To work around this obstacle, Michael Mark Woolfson in Britain proposed in 1964 that the Sun had encountered a 'protostar'. Being less substantial, the protostar would have been more severely influenced by the Sun's gravity than was the reverse case. He envisaged that

**ABOVE** Thomas Chrowder Chamberlin *(University of Wisconsin-Madison)* and Forest Ray Moulton *(University of Chicago)* explored the idea that the planets of the solar system coalesced from material ripped out of the Sun during a close encounter with another star.

**LEFT** Lyman Strong Spitzer. *(Princeton)*

most of the streamer was derived from the interloper which, being cooler, would have more readily condensed. In 1946 Fred Hoyle offered a novel solution, proposing that the Sun had started out as the junior member of a binary star system and its evolved companion had suffered a supernova explosion, leaving behind a cloud of material that was rich in heavy elements and condensed to create the planets.

By the 1940s, however, the tide was turning back to the nebular hypothesis. In 1902 Jeans had calculated the conditions required to cause an interstellar cloud of gas to collapse under self-gravitation. For a given gas density and temperature, he identified a minimum mass.

Carl Friedrich von Weizsäcker in Germany realised in 1944 that a condensing cloud would have been fragmented by turbulence and that each fragment would have made a smaller separate vortex. Hannes Alfvén in Sweden then considered what would have happened to a magnetic field in the collapsing cloud and noted that this would actually have promoted the transfer of angular momentum to the disc and slowed the rate of rotation of the central mass. This avoided the 'flaw' in the original hypothesis. As the temperature of the

contracting gas cloud increased and the atoms were ionised to form an electrically conducting plasma, this would have 'locked in' the magnetic field. As the contraction continued, the intensity of the field would have increased. In effect, the proto-Sun was magnetically coupled to the disc from which the planets later condensed.

In fact, this revival of the nebular hypothesis was based on a discovery made at the end of the 19th century. After Pieter Zeeman's discovery in the 1890s that a magnetic field induced 'splitting' in spectral lines, George Ellery Hale, who set up the Mount Wilson Solar Observatory, realised that sunspots are associated with intense magnetic fields. This prompted him to speculate that the Sun might have a general magnetic field, and in 1939, when the level of sunspot activity was at its 'minimum', he confirmed that the Sun possesses a 'dipole' field. Furthermore, the polarity of the field reverses periodically. On average, the cycle of sunspots peaks every 11 years but the magnetic cycle is twice as long, at 22 years.

The nebular hypothesis was greatly refined in subsequent years. A milestone came in 1969

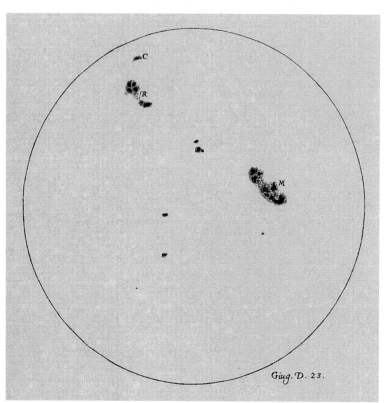

when the Soviet astronomer Victor Sergeevich Safronov published the book *Evolution of the Protoplanetary Cloud and Formation of the Earth and the Planets*. His 'solar nebular disc model' had a lasting influence on how scientists dealt with the formation of planets by the accretion of planetesimals.

This theory was further explored in the early 1990s by George West Wetherill of the Carnegie Institute of Washington, who modelled the process in a computer, plugged in initial assumptions, and let it run to see what happened. Because they express probabilities, these are called Monte Carlo simulations. Slight differences in the initial conditions can rapidly diverge. Wetherill ran many different versions of the same basic model to judge the statistical significance of the results. When he published the results in 1995 he had run over 500 models that varied 20 assumptions to investigate the mass of the proto-Sun, the mass of the disc of gas and dust from which the planets would form, and how giant outer planets affected the formation of smaller, inner rocky ones.

What Wetherill found was that the simulations frequently produced one or more planets with masses similar to Earth in positions where liquid water would be stable on their surfaces. (For the moment, let's use that criterion as a definition of the term 'habitable planet'.) This occurred for stars with masses in the range 0.5 to 1.5 that of the Sun. For stars like the Sun, the formation of at least one habitable planet seemed almost inevitable. In addition, the more massive the disc, the more massive the habitable worlds.

For these initial simulations, Wetherill had 'plugged in' the outer gas giants because his model assumed they already existed. This seemed reasonable because earlier calculations had suggested they would have formed very rapidly, possibly in only a few million years, whereas the inner, rocky planets would have required tens of millions of years to form.

In his initial simulations Wetherill found that if he omitted the gas giants, the model produced more rocky planets than actually exist, extending out almost to where Jupiter now orbits. He concluded that the gravitational influence of mighty Jupiter caused Mars to be smaller than it would otherwise have become,

and also prevented any rocky planet from forming farther out.

But how were the outer, giant planets formed?

In 1972 Alastair Graham Walter Cameron of Harvard developed a model for the formation of giant planets called 'core accretion'. Something similar had been suggested by Gerard Peter Kuiper in Chicago, but Cameron imparted a degree of rigour to the process.

According to this theory, giant planets start off by accreting solid material in much the same way as would produce the rocky planets, but they produce bodies with masses about ten times that of Earth. By that point, these bodies, referred to as 'cores', would have begun to pull in gas from the nebula. It would have been a runaway process, because the larger they became the more readily they would be able to pull in more gas. This would have continued until the proto-star 'switched on' and the remaining nebula was blown back into interstellar space. He reasoned that there would easily have been sufficient time for a planet as large as Jupiter to form.

The core accretion theory gained a lot of support, but Cameron kept looking for other possibilities and in 1978 he suggested an entirely different model. In this scheme, the disc was expected to incorporate pressure waves. In places where the pressure was suddenly increased, gravitational collapse would form 'clumps' that might or might not disperse. Solid material in a stable clump would settle to the centre and accrete to produce a core surrounded by a gaseous envelope. In some cases that body would have sufficient mass to initiate runaway growth.

The advantage of this 'disc instability' model was that it was fast, but it was speculative because the instabilities were difficult to define. Nevertheless, we can observe larger-scale density waves in the spiral arms of galaxies inducing clouds of interstellar gas to collapse in on themselves and create large numbers of young, hot stars, so the process itself is clearly valid. The disc instability model would be very effective in a massive disc, but a disc below a given mass would be unlikely to suffer the requisite instabilities.

As proponents of the core accretion model increased the size of their discs to ensure that by the time the cores had accreted sufficient mass to initiate runaway growth there would still be enough gas present to make the envelopes of the giant planets, the more likely it was that gravitational instabilities in such massive discs would have produced some giant planets anyway.

Perhaps both schemes would operate for discs that were sufficiently massive, with instability rapidly forming the principal member of the star's retinue, in our case Jupiter, and the other giants forming by the slower process of core accretion. All of this activity would have concluded before significant progress was made in forming the smaller rocky planets.

By the end of the 20th century, therefore, we had a fair understanding of how the solar system was formed, and it seemed reasonable to assume that other solar-type stars would possess similar systems of planets.

# Life in the solar system

With the invention of the rocket, astronomers began dreaming of going into space to inspect the planets for themselves.

"Utter bilge," scoffed Richard van der Riet Woolley in 1956. As Astronomer Royal, his voice carried considerable authority.

"Space travel is inevitable," insisted Kenneth William Gatland, a member of the council of the British Interplanetary Society.

The Space Age dawned in October 1957 with the launch by the Soviet Union of the world's first Sputnik. Within a decade it was evident from the results of the first probes to make flybys of Venus and Mars that our investigation of the solar system was undergoing a revolution as profound as that enabled by the invention of the telescope.

At times, astronomers have assumed, usually on philosophical grounds, that various solar system bodies must be inhabited. Let us start in close to the Sun and work outward, reviewing how opinions concerning potential abodes for life have changed over the years.

## Venus

Named after the Roman goddess of love and beauty, the planet Venus shines brilliantly in the sky in the hours after sunset or before sunrise. Its orbit around the Sun has a radius of about 108 million km, which is 0.72 times the radius of the orbit of Earth. The period of its orbit is 225 days. When inspected by telescope, the diameter of the planet was revealed to be only slightly less than that of Earth, so it was widely regarded as a 'twin' of Earth. As such, it seemed likely to be an abode for life.

After investigating Venus from 1877 to 1890 Giovanni Virginio Schiaparelli in Italy concluded the axial rotation of the planet was synchronised with the period of its orbit. Others disagreed, and estimated a variety of periods ranging upward from about 24hr.

In 1915 Charles Edward Housden, a British engineer and amateur astronomer who said the rotation was synchronous, suggested that as a result of keeping one hemisphere facing the Sun and the other in perpetual darkness, the circulation of the atmosphere would be very different to Earth's. Specifically, hot air would rise at the 'subsolar point' and flow at high altitude around to the dark hemisphere, where it would be chilled, descend, and flow back to the illuminated hemisphere at low level. Housden also believed the atmosphere to be wet and predicted there would be an accumulation of ice on the frozen hemisphere.

Thermocouple measurements in the late 1920s by Seth Barnes Nicholson and Edison Pettit using the 100in on Mount Wilson, the largest telescope in the world, revealed that the atmosphere on the dark hemisphere wasn't as cold as it ought to be if the planet's rotation were synchronous. It was concluded from this that the period was several months.

Many of those who believed Venus's rotation wasn't synchronised subscribed to the popular idea, derived from the nebular hypothesis, that the proto-Sun shed a succession of rings that later formed planets; that Mars was an 'ancient' world which had already dried up and 'died'; and that Venus was a 'younger' version of Earth.

Indeed, in 1918 Svante August Arrhenius in Sweden argued Venus was a lush environment resembling Earth as it was several hundred million years ago, when much of its surface was a vast swamp, thick with luxuriant vegetation. In 1924, William Henry Pickering in America suggested that Venus might possess oceans. Water vapour is difficult to identify unambiguously in a planetary atmosphere by spectroscopy owing to its presence in our own atmosphere, but in 1897 the Irish physicist George Johnstone Stoney had announced the atmosphere of Venus to be laden with water vapour, so a hot, wet Venus seemed plausible.

In 1932 Walter Sydney Adams and Theodore Dunham of the Mount Wilson Observatory noticed absorption bands in the infrared that were later attributed to carbon dioxide. The presence of a 'heavy' gas in the upper atmosphere meant that the lower atmosphere would be rich in carbon dioxide, and since carbon dioxide is a 'greenhouse' gas it followed that the surface would be rather warmer than the planet's location close to the Sun would imply. In 1939 Rupert Wildt at Princeton argued that the surface temperature likely exceeded the boiling point of water, in which case there could not be oceans.

But in 1929 Bernard Ferdinand Lyot of the Meudon Observatory in Paris had reported a polarisation study that attributed the brightness of the clouds of Venus to their being composed of small droplets. Presuming the droplets to be water, in 1955 Fred Lawrence Whipple and Donald Howard Menzel at Harvard University argued that in a predominantly carbon dioxide atmosphere the water cycle must be so vigorous that erosion by carbonic acid, created by water absorbing carbon dioxide, must have transformed the planet into a global-scale ocean.

Furthermore, the ever imaginative Fred Hoyle in England claimed in 1955 that the

fluid in Venus's ocean was not water but hydrocarbons (oil). It was his view that there could not be any water left. The cloud layer, he said, was a dense smog made up of dust motes and tiny droplets of oil. His logic was that if hydrocarbons that accumulated at shallow depth in the crust eventually broached the surface to create lakes, then the volatiles would evaporate and be oxidised, thereby drawing oxygen from the atmosphere. The process would continue, he said, until either all of the oil was oxidised or all of the surface water had evaporated, terminating the liberation of oxygen by the dissociation of water vapour in the upper atmosphere by solar ultraviolet. The presence of the smoggy clouds indicated the oxygen had run out first.

Only when radio astronomers developed a sensitive radiometer in 1956 was it possible to directly investigate the surface of the planet in terms of its microwave emissions. However, a temperature exceeding 300°C was puzzling. Even though the energy in sunlight received by Venus is twice that at Earth's distance from the Sun, the brilliant white clouds in the upper atmosphere reflect 85% of it back into space.

When a radio telescope was operated as a radar, it used the Doppler effect to determine that Venus rotates extremely slowly. What is more, it spins 'in reverse'. In the vernacular of astronomers, this is retrograde rotation. In

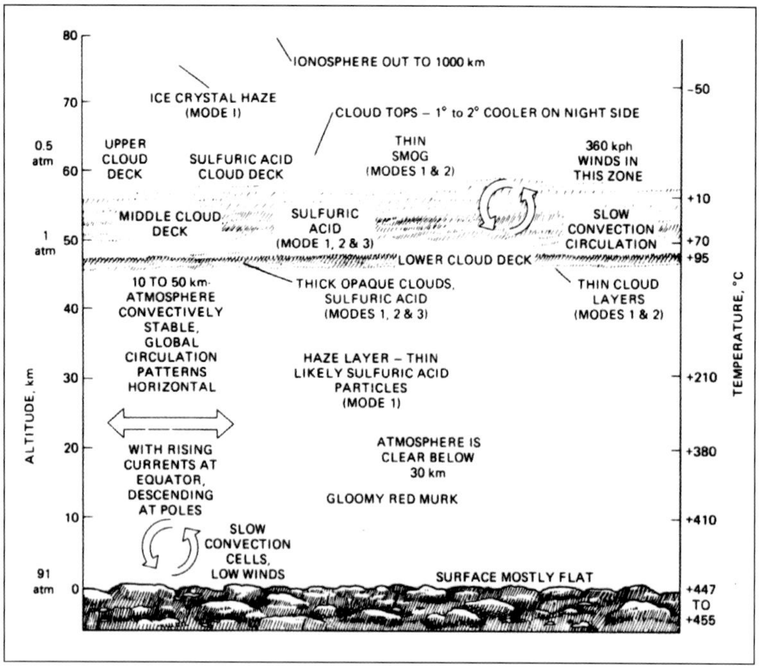
80
70
0.5 atm 60
1 atm 50
40
30
20
10
91 atm 0
ALTITUDE, km
IONOSPHERE OUT TO 1000 km
ICE CRYSTAL HAZE (MODE I)
CLOUD TOPS – 1° to 2° COOLER ON NIGHT SIDE
UPPER CLOUD DECK
SULFURIC ACID CLOUD DECK
THIN SMOG (MODES 1 & 2)
360 kph WINDS IN THIS ZONE
MIDDLE CLOUD DECK
SULFURIC ACID (MODE 1, 2 & 3)
SLOW CONVECTION CIRCULATION
LOWER CLOUD DECK
10 TO 50 km ATMOSPHERE CONVECTIVELY STABLE, GLOBAL CIRCULATION PATTERNS HORIZONTAL
THICK OPAQUE CLOUDS, SULFURIC ACID (MODES 1, 2 & 3)
THIN CLOUD LAYERS (MODES 1 & 2)
HAZE LAYER – THIN LIKELY SULFURIC ACID PARTICLES (MODE 1)
WITH RISING CURRENTS AT EQUATOR, DESCENDING AT POLES
ATMOSPHERE IS CLEAR BELOW 30 km
GLOOMY RED MURK
SLOW CONVECTION CELLS, LOW WINDS
SURFACE MOSTLY FLAT
-50
+10
+70
+95
+210
+380
+410
+447 TO +455
TEMPERATURE, °C

**ABOVE The structure of the atmosphere of Venus, as derived from the data provided by the Pioneer-Venus atmospheric penetration probes in 1978. The hot lower atmosphere is fairly stagnant, with the main action occurring at high altitude; indeed, the base of the cloud deck lies at an altitude of almost 50km. The droplets which form the clouds above that level are not water (as on Earth) but aerosols of sulphuric acid.** *(NASA/AMES)*

effect, the planet is 'upside down'. Interestingly, at 243 days, Venus takes longer to spin on its axis than it does to travel around the Sun.

In December 1962 the first spacecraft to make a flyby of Venus established it to have an extremely hot surface, at about 480°C. This ruled out both the 'global ocean' and 'lush swamp' theories.

The first in-situ observations were made by a Soviet 'entry probe' in October 1967. It descended in the equatorial zone in darkness, with a line of sight to Earth, and when it fell silent the temperature was 280°C and the ambient pressure was 22 bars. When a NASA spacecraft flew by the following day its trajectory crossed behind the planet as seen from Earth. The manner in which the radio signal was attenuated by the planet's atmosphere yielded a profile from which extrapolation indicated the pressure at the surface to be in the range 75 to 100 bars, so when the Soviet probe ceased transmitting it was still 27km above the surface.

A reinforced Soviet probe reached the surface of Venus intact in July 1972. It came down just after local dawn, and functioned perfectly. The chemical analyser established the atmospheric composition as 96% carbon dioxide, 3% nitrogen, and at most 0.1% oxygen. The wind speeds decreased from 100m/sec at an altitude of 49km, which marked the base of the cloud deck, to less than 1m/sec at a height of 10km, meaning that the extremely dense lower atmosphere must be stagnant. The conditions at the surface were 470(±8)°C and 90(±2) bars.

In 1969, a study of the index of refractivity of the atmosphere had established that the clouds were not droplets of benign water but aerosols of sulphuric acid formed by photochemical oxidation at altitudes exceeding 60km. Oxygen released by the dissociation of carbon dioxide oxidises sulphur dioxide to sulphur trioxide, which the process of hydration transforms into droplets of sulphuric acid. As the droplets 'rain out' they are thermally disrupted on reaching the base of the cloud deck, where the temperature is 100°C. They liberate sulphur trioxide which, upon encountering carbon monoxide, regenerates carbon dioxide to complete the cycle. On Earth, the cycle of precipitation ranges from the surface to an altitude of about 18km. On Venus, it is isolated from the surface and operates entirely in the upper atmosphere where the temperatures are more moderate.

Far from being a likely 'twin' for Earth, Venus was revealed to be a veritable Dante's *Inferno* that held out no prospect as a possible abode for life. Or at least on the surface. There remains just a chance that conditions might be conducive in the cool upper atmosphere.

## Mars

Shining with a 'blood red' appearance, the planet Mars was named after the Roman god of war. Owing to its distinctive hue, it is often referred to as the 'Red Planet'.

Its orbit around the Sun is elliptical, being 206 million km at its closest and 249 million km at its farthest, giving it a mean heliocentric distance of about 1.5 times the radius of Earth's orbit. In terms of our calendar, it takes 687 days to make one circuit of the Sun.

In the 17th century, astronomers armed with early telescopes saw markings on the surface of Mars. Careful study enabled them to determine that it turned on its axis in a period of just over 24hr. They later noted that its axis was tilted by an angle comparable to that of Earth's axis. This tilt, together with the elliptical orbit, gave rise to seasons that caused the white polar caps to wax and wane. In the 19th century,

astronomers with better telescopes noted seasonal variations which they suspected might indicate vegetation.

A milestone in the study of Mars was when Giovanni Virginio Schiaparelli in Milan conducted a trigonometric survey when the planet was very well presented for viewing in 1877. His sketches depicted networks of fine lines that he described as 'canali', meaning channels. But they were visible only when a condition known to astronomers as 'seeing' was favourable. Schiaparelli's report was received with some scepticism.

Percival Lowell, a wealthy Bostonian with a passion for astronomy, was eager to investigate. He built an observatory at Flagstaff in the Arizona Territory which, by virtue of being on a plateau at an elevation of 7,200ft, had excellent seeing for a large part of the year. He installed a borrowed 18in refractor and was ready to start observing when Mars was once again favourably presented for viewing in 1894. As Lowell would delightfully write later, the canali were visible "hour after hour, day after day, month after month".

Back in Boston, Lowell reviewed almost 1,000 drawings and decided that the canali

**ABOVE The diameter of Mars is just over half that of Earth.** *(NASA)*

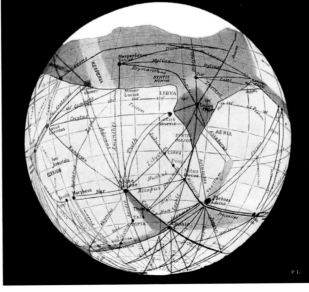

**LEFT Percival Lowell and his vision of Mars crisscrossed by canals made by the natives to transport water from the polar caps to oases in the deserts.** *(Photograph from Lowell Observatory. Drawing from Lowell's 1908 book* Mars as the Abode of Life*)*

were laid out purposefully. He published his conclusions in late 1895 in a volume that he entitled simply *Mars*: "Firstly, that the broad physical conditions of the planet are not antagonistic to some form of life; secondly, that there is an apparent dearth of water on the planet's surface and, therefore, if beings of sufficient intelligence inhabit it, they would have to resort to irrigation to support life; thirdly, that there turns out to be a network of markings covering the disc precisely counterparting what a system of irrigation would look like; fourthly, and lastly, there is a set of spots placed where we should expect to find the land thus artificially fertilised, and behaving as such constructed oases should." Thus the Martians had established themselves where the canali intersected on the ochre tracts, at the dark spots which he called oases.

Lowell's portrayal of Mars as a dying world was so evocative that it inspired Englishman Herbert George Wells to write a story called *The War of the Worlds*, in which the Martians invade Earth, tempted by its water. Following serialisation in 1897, it was issued as a book in 1898 and became an instant best-seller.

In reality, Lowell's theory faced a major problem concerning conditions at the surface of Mars. In his opinion the temperature compared favourably with that of a summer's day in the south of England, with sufficient air pressure for water to flow efficiently along the canals. But the English naturalist Alfred Russel Wallace wrote in a rebuttal, "All physicists, are agreed that, owing to the distance of Mars from the Sun, it would have a mean temperature of about -35°F even if it had an atmosphere as dense as ours. But the very low temperatures on Earth at the equator, at a height where the barometer stands at about three times as high as on Mars, proves, that from the scantiness of atmosphere alone Mars cannot possibly have a temperature as high as the freezing point of water," which, he insisted, was "wholly incompatible with the existence of animal life". Wallace's conclusion was that Mars "is not only uninhabited by intelligent beings such as Mr. Lowell postulates, but is absolutely *uninhabitable*". In fact, Wallace erred in calculating the temperature on Mars but his analysis was still superior to that of Lowell and his conclusion stands.

There were numerous spectroscopic efforts to measure the composition of the atmosphere and its surface pressure, but it wasn't straightforward and the results conflicted. With estimates of about 100 millibars, it was reasoned that conditions might just be tolerable for a type of life equivalent to lichen. This isn't a plant – it is a symbiotic association of a fungus and an alga, with the fungus providing an isolated environment that protects the alga and lives off the wastes of the alga's photosynthesis. Lichens can survive in places that neither the fungus nor the alga alone could live. Because algae and fungi are more primitive than 'higher' plants, it seemed reasonable that they might have developed independently on different worlds.

But in 1964 an analysis by Hyron Spinrad of the Jet Propulsion Laboratory of the California

Institute of Technology put an upper limit on the pressure at the surface of 25 millibars, and put the partial pressure of carbon dioxide at between 4 and 5 millibars. The remainder of the atmosphere was presumed to be primarily molecular nitrogen, but it would be difficult to detect. Harold Clayton Urey of the University of California at San Diego had made the radical suggestion a few years earlier that nitrogen was absent. If true, the atmosphere must be composed almost entirely of carbon dioxide and have a pressure of only several millibars, in which case water would not be stable on the surface.

In July 1965 the first spacecraft to fly by Mars revealed it to possess an ancient cratered surface and an atmosphere of at least 95% carbon dioxide with a surface pressure in the range 4 to 6 millibars. Despite these shocking results, NASA went ahead with a mission designed to seek life on the planet.

The two Viking landers that tested for life on Mars in 1976 used methods that were recommended in 1965 by a panel of experts led by the renowned molecular biologist Joshua Lederberg. They said it would be reasonable to presume that life originated independently on the planet, but pointed out that whilst if there were plants there would certainly be microbes, there might only be microbes. Hence the tests should be directed at microbial life. In short, as the report stated, it would be necessary to "assume an Earth-like carbon-water type of biochemistry as the most likely basis of any Martian life".

In view of how biological cells function, one strategy was to seek evidence of cellular reproduction but this was a discontinuous process, the rate of which was variable from one species to the next and even in different conditions for a given species. That would make employing it as a test very difficult in the context of an exotic environment. As an ongoing process that could be measured in a number of ways, for example by changes in acidity or the evolution of gases, metabolism was more readily testable and was more likely to produce a definitive result. The report urged a multifaceted test because "no single criterion is fully satisfactory, especially in the interpretation of negative results".

The results from sites on opposite sides of the planet were disputed. Gilbert Levin, principal

LEFT Joshua Lederberg circa 1962.

BELOW Clockwise from top left, the principal investigators of the biology experiments conducted by the Viking landers on Mars: Harold P. Klein *(NASA/AMES)*, Vance I. Oyama *(NASA/AMES)*, Gilbert V. Levin *(Spherix)*, Norman H. Horowitz *(Caltech/James McClanahan)*.

investigator for one of the experiments argued, "The accretion of evidence has been more compatible with biology than with chemistry – each new test result has made it more difficult to come up with a chemical explanation, but each new result has continued to allow for biology." If a terrestrial sample had given the results they had seen on Mars, "We'd unhesitatingly have described [it] as biological." But Vance I. Oyama, heading up another experiment was sceptical, "There was no *need* to invoke biological processes." Norman H. Horowitz, head of the remaining experiment agreed, but pointed out it was "impossible to prove that any of the reactions … were *not* biological in origin".

So prior to the Viking landings no one knew whether there was life on Mars, and, sadly, no one knew afterwards either!

The biology team leader Harold P. Klein later concluded that the assumption that Martian microbes must resemble terrestrial ones ought to be dismissed, and scientists should review whether the Viking data offered any clues as to "whether there might be some less obvious kind of life on Mars".

In fact, even as the Viking landers were seeking carbon-synthesising microbes on Mars, biologists on Earth were discovering the first examples of an entirely new class of microbial life.

**BELOW A 'black smoker' some 10m tall discovered by EV *Nautilus* on the ocean floor near the Galápagos Islands. The superheated water in the hydrothermal vent is teeming with strange life forms.** *(Ocean Exploration Trust)*

## Terrestrial extremophiles

Earth accreted from the solar nebula some 4.5 billion years ago, and within around 100 million years it had cooled sufficiently for a hydrosphere to form. At that time the surface of the planet was dominated by volcanism that pumped up the atmosphere with carbon dioxide, creating a 'greenhouse' which trapped solar energy.

The oldest known rocks are around 4 billion years old, but these have been altered by subsequent processing and so can reveal nothing of the likelihood of life at that time. The first convincing evidence is in well-preserved 3.5 billion-year-old rocks at Barberton in South Africa and in the Pilbara in Australia that contain various indicators of microbial ecosystems.

It had been believed that water in the hottest of geothermal springs must be sterile, but in the early 1970s microbes were discovered living in water at 85°C in Yellowstone National Park.

Then in 1977 a submersible which was investigating the rift in the Galápagos Ridge discovered a hydrothermal vent on the ocean floor that was spewing out a super-heated plume of water. A similar vent located in 1979 on the East Pacific Rise was so rich in dissolved minerals that a 'chimney' had formed. Such 'black smokers' supported local ecosystems that hosted many species of life. Next, some well-preserved mineral textures indistinguishable from smokers were found in a 3.26 billion-year-old sediment in Australia.

In 1982 it was realised that these isolated food chains were based on single-celled organisms that derived their energy from the nutrients of the hydrothermal vent. These thermophilic microbes were the first known members of a new class of life named archea (the old ones).

Biological cells are autotrophs or heterotrophs (or perhaps both) in that they either manufacture organic compounds themselves from raw materials such as carbon dioxide (autotrophs, meaning self-feeders) or they draw organic material from their environment and process it into whatever they require (heterotrophs). Which type came first was disputed.

The heterotrophic origin theory was proposed in the 1920s by the Russian chemist Alexander Ivanovich Oparin, and appeared reasonable. After all, why should not the earliest

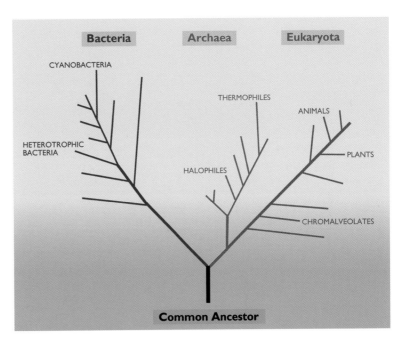

Bacteria  Archaea  Eukaryota

CYANOBACTERIA

THERMOPHILES

ANIMALS

HETEROTROPHIC BACTERIA

PLANTS

HALOPHILES

CHROMALVEOLATES

**Common Ancestor**

life have exploited the organic compounds present in its environment? But opinion now favours an autotrophic origin. This reasoning was reinforced by the discovery of yet another unusual microbial ecosystem.

In the mid-1980s microbes were found in the interstices between grains of rock at a depth of 1km beneath the surface of the Piceance Basin in Colorado. There were some heterotrophs present which consumed the remains of plant detritus bound up in sedimentary rock. They were similar to those existing on the surface, but had adapted to the hot anoxic environment by remaining static. Most of the microbes were autotrophs that exploited the heat and hydrocarbons which were toxic to 'conventional' life. These microbes were named lithotrophs because they 'lived off rock'.

Autotrophs are well represented among the archea. Those which thrive in a hydrothermal environment that is lacking both oxygen and light are collectively known as either anaerobic autotrophs or chemolithotrophs. Many require only water enriched with volcanic gases and nutrients.

The fact that none of the archean autotrophs use sunlight implies the process of photosynthesis developed later. The ability to draw energy from sunlight was a major advance because it is more efficient.

Unrestricted to hydrothermal vents, a single-celled cyanobacteria that could use chlorophyll for photosynthesis was free to colonise the planet. They initially formed thin mats on the floors of shallow seas. Layers of rock were accreted as the mats trapped particulates in suspension and minerals that precipitated out of the water. These colonies are called stromatolites. There is disputed evidence for fossil stromatolites in the Pilbara, but no doubt regarding the biological origin of 3.5 billion-year-old fossils at Baffin Island in Canada because they can be matched, point for point, with the 'living fossils' that still survive in the highly saline Shark Bay of Western Australia.

The development of archea on the early Earth had profound implications for life elsewhere.

## Current efforts to seek life on Mars

When NASA returned to Mars with orbiters and landers in the final years of the 20th century and the early years of the 21st, its 'follow the water' strategy was designed to seek geological indications of the presence of surface water in order to determine whether conditions had ever been conducive to the development of life. Another search for current life was not attempted until the British Beagle 2 lander in 2003. This

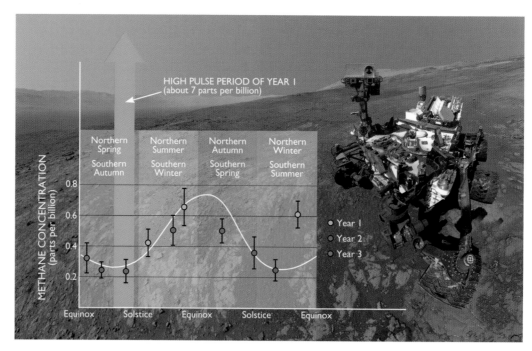

RIGHT In addition to a major spike in the concentration of methane in the atmosphere of Mars, the Curiosity rover has recorded year-on-year cyclical variations. *(Data from NASA/JPL-Caltech/graphic W. D. Woods)*

BELOW Jupiter's moon Io, here observed by the Galileo mission, is the most volcanically active body in the solar system. Its vents hurl plumes to heights of hundreds of kilometres, and when this material falls back it produces colourful surface deposits. *(NASA/JPL-Caltech/University of Arizona)*

was delivered by the European Space Agency's Mars Express mission. It appears to have landed safely, but then malfunctioned. It was to have employed a 'mole' to burrow to a depth of at least 1m to collect a sample in what was expected to be a more benign environment than the harsh surface.

The Curiosity rover that NASA landed in 2012 explored the bed of an ancient lake which could have been friendly to microbes. It then set off up the flanks of a hill to study its stratification and determine how the early habitable environment became more stressful for any microbes that may have been present.

Using a similar rover, to be launched in 2020, NASA intends to seek samples at another site to assess the potential for preserved 'biomarkers' proving that life once existed.

Meanwhile, the European Trace Gas Orbiter, which settled into orbit around the planet in 2016, is to scan the atmosphere to create profiles from the surface up to 160km to determine spatial and temporal variations of methane and other trace gases and hopefully localise their sources. If the methane is found in the presence of either propane or ethane, this combination will strongly imply a contemporary biological origin.

## The moons of Jupiter

When the first Voyager spacecraft flew through the Jovian system in March 1979 it revealed

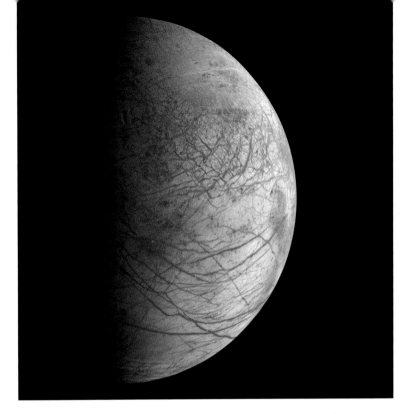

RIGHT **A view of the Jovian moon Europa by the Galileo spacecraft showing long cracks in its icy shell.** *(NASA/JPL-Caltech)*

each of the four large moons to possess its own distinct character.

The innermost, Io, stole the show with active volcanoes that issued plumes of material which fell back to 'paint' the surrounding surface. Although this came as a surprise, it oughtn't to have. In *Science* that very week, Stanton Jerrold Peale, Patrick M. Cassen and Ray T. Reynolds, researchers at the NASA Ames Research Center in California, predicted the moon might display widespread volcanism.

This arises because Io's axial rotation is synchronised with the 42.5hr period of its orbit around Jupiter, and it is so close to the planet that there is a 100m high bulge on the hemisphere which permanently faces the planet. In addition, cyclical perturbations from the next moon out, Europa, make Io's orbit slightly elliptical. The fluctuation in the gravitational field induces tidal stresses on the bulge, and on the interior of the moon. The physical stress manifests itself as heat. This has differentiated the moon into a central core, a fluid mantle, and a silicate crust, and is now driving volcanism.

Similar but weaker gravitational stresses from orbital resonances with Io and the larger but farther out Ganymede are heating the interior of Europa. However, in this case, as the Voyagers revealed, this moon is completely covered with ice. During its tour of the Jovian moons, the Galileo spacecraft showed the ice to be cracked in patterns which could occur only if the surface were a thin shell entirely isolated from the rocky interior. This, in turn, implied the existence of a subsurface ocean. Some areas suggested plumes of warm water had risen up and temporarily weakened the ice. Indeed, in other areas it seemed that patches of open water had enabled blocks of ice to jostle one another prior to the refreezing of the exposed fluid.

During close flybys, the readings by the magnetometer established Europa to possess a magnetic field whose characteristics indicated a salty ocean. The axis of Jupiter's magnetic field is tilted about 10° to the spin axis of the

BELOW **There is an ocean beneath the icy shell of Europa. By analogy with 'black smokers' at hydrothermal vents on the floor of the terrestrial oceans, life may have developed on this Jovian moon.** *(NASA/K. P. Hand et al)*

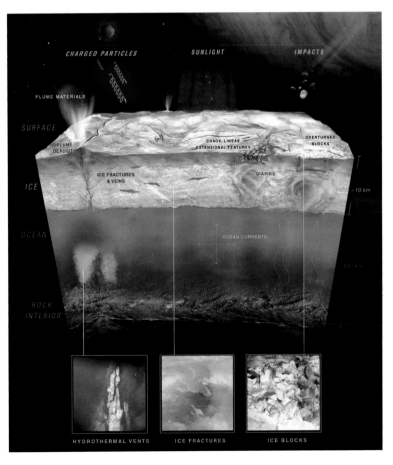

ABOVE Ganymede, the largest of the Jovian moons, as viewed by the Galileo spacecraft. (NASA/JPL-Caltech)

planet and thus to the plane in which the main moons orbit. Also, the field rotates with the planet in a period of only 10hrs, much faster than the moons travel around the planet. Thus sometimes the moons are north of the magnetic equator and other times they are

BELOW The internal structure of Ganymede inferred from measurements by its magnetometer and from radio tracking of the Galileo spacecraft during close flybys. (Data from NASA/JPL-Caltech/graphic adapted by W. D. Woods from Wikipedia/kelvinsong)

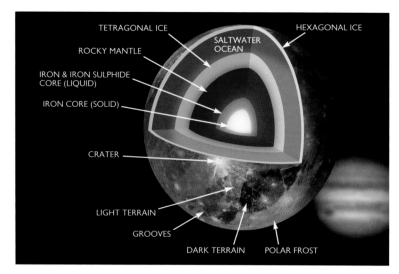

south of it. As a result, the ambient field in Europa's location periodically reverses its polarity. If Europa's weak magnetic field were produced by electrical currents induced by the changing ambient magnetic field, then the moon's magnetic poles ought to be near the equator and they should migrate. Galileo found this to be the case. This was strong evidence for an ocean of salty water beneath the shell of ice. In addition, precise radio tracking of how the spacecraft's trajectory was deflected while making close flybys supported this conclusion. Calculations have suggested the shell is very shallow, possibly only several kilometres thick. It caps a layer of water 100km deep. In fact, the moon has more liquid water than exists in all the oceans of Earth!

This raised the question of whether the heat caused by gravitational stresses would produce hydrothermal vents on the silicate floor of this ocean, just like the ones on volcanically active areas of the ocean floor on Earth. If so, why shouldn't extremophiles equivalent to terrestrial archea have independently originated and occupy that niche environment?

In terms of its diameter, Ganymede is not only the largest moon in the Jovian system, it is the largest moon in the solar system. Indeed, the fact that it is larger than the planet Mercury indicates it is a small planet that happens to be orbiting Jupiter. The Voyagers showed its surface to combine large dark cratered regions and brighter terrains of narrow, parallel grooves. The Galileo spacecraft revealed the surface in greater detail.

In addition, the tracking data showed the moon to be fully differentiated into an icy crust 100km thick over 800km of 'soft ice', a mantle of silicate, and an iron-rich liquid core. The chemical differentiation required to create such a core meant that the moon had undergone significant internal heating. For a body as large as Ganymede, the accretional and radiogenic heating would have been sufficient to produce partial melting. This may have been further stimulated by a pulse of heat from gravitational stresses as the moon's rotation was slowed to synchronise with its orbital period. After that, orbital resonances with Europa might have kept the interior warm. The magnetometer found that the core generates a dipole magnetic field of sufficient strength to award Ganymede its own magnetosphere and polar aurorae.

After it was discovered that Jupiter's magnetic field was inducing electrical currents in Europa, the Galileo magnetometer team re-examined their Ganymede data for similar secondary effects superimposed upon the moon's intrinsic dipole field. They found an induced field which was an order of magnitude weaker than the intrinsic one, and was varying in the same way as the induced field of Europa. Hence the thick outer layer of Ganymede is a combination of an icy crust and a substantial ocean of electrically conducting fluid.

In the case of Callisto, the outermost of the large moons, the Galileo tracking data implied that the interior isn't fully differentiated. It seems to be a mixture of rock and ice, with the proportion of rock increasing with depth. This situation has been called 'half baked'. The magnetometer indicated that something inside was conducting an induced electrical field. This generated a weak magnetic field that interacted with the ambient field and varied its direction in the same manner as Europa. However, although the conducting fluid may be global in extent, it must be fairly deep beneath the surface, perhaps several hundred kilometres, as there is no evidence at the cratered surface to suggest its presence. In fact, it might be only a zone of slushy ice. Being so far from Jupiter, and thus free of orbital resonances, the tidal stresses on Callisto are weak. The heat to maintain water in a liquid state probably derives from the decay of radioactive elements present in the rock.

### Future Jupiter missions

Several projects are now underway to further investigate the ocean moons of Jupiter.

In 2012 the European Space Agency initiated development of the Jupiter Icy Moon Explorer (JUICE) mission. It is to launch in 2022 and reach Jupiter in 2030. After multiple flybys of Callisto and Europa, it will enter orbit around Ganymede for a close-up science mission.

In 2015 NASA instructed its own Jet Propulsion Laboratory and the Applied Physics Laboratory of Johns Hopkins University to undertake the Europa Clipper mission to investigate the habitability of Europa and help in the selection of sites for a future lander. The

spacecraft is to remain in orbit around Jupiter and make a series of very close flybys of the moon using an ice-penetrating radar, short-wave infrared spectrometer, topographical imager, and mass spectrometer. The lander concept is still under study. A particularly adventurous option would be to have a probe melt its way through the icy shell in order to explore the seabed in search of life.

## The moons of Saturn

During its passage through the Saturnian system in 1980, Voyager 1 was able to show that Enceladus had a very highly reflective surface but it did not fly close enough to examine it in detail; this was done a year later by Voyager 2.

With Mimas and Enceladus being similar in size and location, close to Saturn, it had been presumed that they would be twins. But whereas Mimas was entirely cratered, a large portion of the surface of Enceladus was found to have undergone extensive resurfacing by a cryovolcanic process. The sparsity of craters on these plains was taken to indicate their relative youth. In addition, some sites suggested that fluid had oozed from fractures and solidified on the surface. The extent of the resurfacing and tectonic activity on this small moon was

**BELOW The surface of Mimas, the small inner moon of the Saturnian system, is impact scarred, in contrast to the similarly sized Enceladus which orbits the planet a little farther out. This image was obtained by the Cassini spacecraft.** *(NASA/JPL-Caltech/ Space Science Institute)*

ABOVE The 'tiger stripes' in the southern polar region of Enceladus imaged by the Cassini spacecraft. They mark fractures in the icy shell where material is 'jetting' from the interior. *(NASA/JPL-Caltech/ESA/Space Science Institute)*

a considerable surprise. Little further progress was possible until the Cassini mission entered orbit around Saturn.

In January 2005 Cassini imaged Enceladus from 367,000km and revealed the leading hemisphere (not seen by the Voyagers) to be an extraordinary landscape of swirling, curvilinear wrinkles. When it passed within 1,264km in February, the smooth terrain on the southern part of the trailing hemisphere was shown to bear faults, fractures, folds and troughs. On a 500km flyby in March, Cassini identified a tenuous envelope. On both close flybys, the magnetometer noted that Saturn's magnetic field was 'bent' around the moon by electrical currents generated by the interaction of the magnetosphere with neutral atoms of gas. In addition, when neutrals were ionised by plasma, they were 'picked up' by the magnetic field and made to oscillate at frequencies that enabled them to be identified as $O^+$, $OH^+$ and $H_2O^+$ ions. This indicated the presence of water vapour. The gravity of Enceladus is so weak that its escape velocity is a mere 212m/sec. As gas would readily leak away, there would be a net motion outward. The fact that the envelope persisted implied a source of replenishment. It was decided to lower the altitude of the next flyby to 175km in order to investigate the situation further.

When Cassini returned in July 2005 it penetrated the electrically conducting envelope, which the magnetometer revealed to be concentrated at the south pole. At an altitude of 270km, just as the trajectory attained its most southerly latitude, an instrument saw a peak in the number of particles striking the spacecraft and a mass spectrometer measured a large peak in the abundance of water vapour. The variation of vapour density with altitude implied there was a localised source on the surface.

In addition to the bright rings with which any telescopic observer is familiar, Saturn has a number of fainter ones. The most concentrated part of the diffuse 'E' ring is close to the orbit

LEFT While orbiting Saturn, the Cassini spacecraft discovered plumes of material emanating from Enceladus. This is a false-colour display. *(NASA/JPL-Caltech/Space Science Institute)*

of Enceladus. In flying through this ring, Cassini found it to be made of water-ice grains. An infrared spectrometer on the vehicle provided the clue to the source of this material.

The surface of Enceladus is the coldest in the Saturnian system because it has a very bright surface that reflects sunlight. The temperature at the south pole was found to be the warmest place on the moon. Optical imagery showed a number of arcuate features dubbed 'tiger stripes'. Although the average temperature across the south pole was just 85K, it rose to 145K in the 'stripes'. When a re-analysis of imagery taken in January showed the moon as a crescent with a glow around its south polar region, it was decided to look at this again in November 2005. The new images showed 'jets' from many sources rising 500km into space (twice the radius of the moon). Evidently there was liquid water at shallow depth, and the 'stripes', running parallel, 40km apart, for a distance of 140km, marked fractures through which the liquid made its way to the surface and was vented, with the stream of vapour carrying particles of ice. Some of this material fell back as a bright 'snow' (explaining the very high albedo of the surface) but the rest escaped to replenish the 'E' ring.

Direct sampling on later flybys indicated the source to be a salty subsurface ocean. Measurements of how the moon 'wobbles' as it orbits Saturn, a behaviour called libration, implied that beneath a detached shell of ice there is a global ocean that is perhaps 10km deep and is in contact with the rocky interior. The analysers on Cassini also detected trace amounts of simple hydrocarbons such as methane, propane, acetylene, and formaldehyde in the plumes. In fact, the composition was similar to that observed in most comets.

The intensity of the activity varies significantly as a function of the position of Enceladus in its orbit around Saturn, with the plumes being four times brighter when it is at its farthest from the planet than when it is at its nearest. The fissures are under compression (shut) when the moon is closest to Saturn, and then under tension (open) when farthest from the planet.

Due to its high ratio of surface area to volume, Enceladus must have rapidly lost

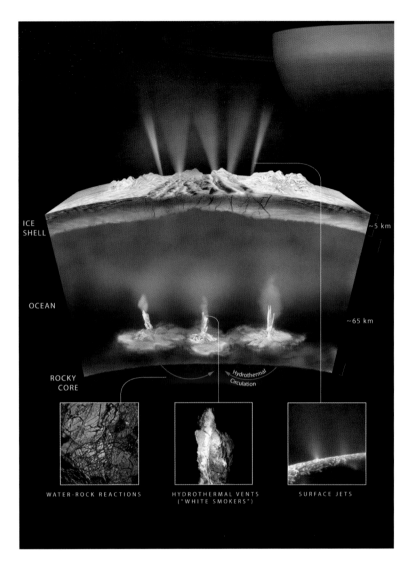

ICE SHELL

~5 km

OCEAN

~65 km

ROCKY CORE

*Hydrothermal Circulation*

WATER-ROCK REACTIONS

HYDROTHERMAL VENTS ("WHITE SMOKERS")

SURFACE JETS

its accretional heat but radiogenic heating could have produced a warm core and a subsurface ocean, the latter particularly if there was a significant amount of methane and/or ammonium hydrate in the ice. The ocean is being maintained by the heat from gravitational tides, most likely a resonance with Dione, which orbits farther out.

Although Enceladus may appear to be just another moon that has an internal ocean, it differs from those in the Jovian system because it has a 'hot spot' that has fractured the icy shell and is driving the jet activity. All else being equal, the heat source could well make that crucial difference between life developing and it not.

Titan, the first member of Saturn's retinue to be discovered, was spotted by Christiaan Huygens in 1655. By following it from night to night, he measured its orbital period as 16 days.

**ABOVE** There is an ocean beneath the icy shell of Enceladus. It is possible that life has developed around hydrothermal vents on the silicate crust. *(NASA/JPL-Caltech/ SwRI)*

It is the only one of Saturn's moons that is large enough to show a disc, but efforts to measure its diameter showed considerable variation. Nevertheless, it was apparent early on that it was comparable in size to Jupiter's Ganymede.

There were occasional reports of extremely faint markings upon its distinctly orangey disc. Bernard Ferdinand Lyot of the Pic du Midi Observatory in France made a lengthy study of it, and drew a variety of light and dark bands and large splotches.

In 1942, Gerard Peter Kuiper initiated a spectroscopic study of Titan using the 2.1m reflector of the McDonald Observatory on Mount Locke in Texas, reporting evidence for a substantial atmosphere. A later study established the presence of methane.

A major objective for Voyager 1 in the Saturnian system was a Titan flyby that would not only perform remote sensing but also pass behind the moon as viewed from Earth so that the manner in which the radio signal was modified by passing through the atmosphere of the moon would yield insight into its composition and physical characteristics.

Titan proved to be completely enshrouded with cloud. Intriguingly, there was no observable weather system. However, limb views showed layering believed to be a stack of photochemical hazes. The atmosphere was rather more substantial than predicted, having a surface pressure of 1.5 bars. The primary constituent was molecular nitrogen but the orange haze meant there was another constituent. The spectroscopic data indicated assorted carbon-nitrogen compounds, including the hydrocarbons ethane, acetylene, and ethylene.

At 5,150km in diameter Titan is almost as large as Jupiter's Ganymede, but its lower bulk density implies it is half rock and half ice. However, having formed farther out in the solar nebula, Titan would have acquired volatiles that wouldn't have been present at the heliocentric

**BELOW** The Otto Struve Telescope of the McDonald Observatory on Mount Locke, Texas, has a mirror diameter of 2.1m. *(McDonald Observatory)*

distance where the moons of Jupiter formed. It was reasoned that Titan would have had sufficient radioactive elements for its interior to become thermally differentiated, with the resulting release of volatiles to generate an early atmosphere of methane (carbon drawn from the solar nebula) and ammonia (nitrogen). Once the ammonia was dissociated by the ultraviolet in sunlight, the hydrogen escaped and the nitrogen was retained. Nowadays, when methane in the upper atmosphere is dissociated, the highly reactive ions, known as free radicals, soon recombine to produce a wide variety of organic compounds.

Titan's surface temperature was measured at 93K, with a variation of only a few degrees from pole to pole. Because the temperature was near the triple points of ethane and methane, these would precipitate as rain. In the mid-1990s, infrared observations from Earth detected variations that were suggestive of clouds in the lower atmosphere, supporting the notion of a rain of hydrocarbons.

It was speculated that a hydrocarbon rain might even flow across the surface and drain into hydrocarbon seas. It was calculated that the eccentricity of Titan's orbit around Saturn would raise tides of several metres amplitude. Over time, the argument said, exposed terrain would be physically and chemically eroded away, creating a predominantly oceanic landscape.

After entering orbit around Saturn in 2004, the Cassini spacecraft released the Huygens probe on a trajectory that would penetrate the atmosphere of Titan. The probe was to report on conditions during a parachute descent and hopefully also on the nature of the surface. It was designed to float, just in case it were to splash down.

As Huygens descended on its parachute in January 2005 and dropped out the clouds its imagery showed a landscape resembling river valleys on Earth, but the liquid couldn't have been water. It landed on a low-lying plain that looked as if it might have been regularly flooded by run-off.

**ABOVE** When Voyager 1 flew by Titan in 1980 it could see only an orange disc because hydrocarbons in the atmosphere made it opaque (top left, *NASA/JPL-Caltech*). The broader spectral range of the camera of the Cassini spacecraft combined visible and infrared and could map the moon (lower left, *NASA/JPL-Caltech/University of Arizona*). After being release by Cassini, the Huygens probe touched down on Titan in 2005 by parachute. The depiction by an ESA artist is based on the information provided by the probe.

The Cassini orbiter was equipped with a radar instrument capable of imaging the surface of Titan through the haze. Although it was revealed not to be an ocean world, its landforms included expanses of exposed ice and large fields of dunes, possibly of particulate

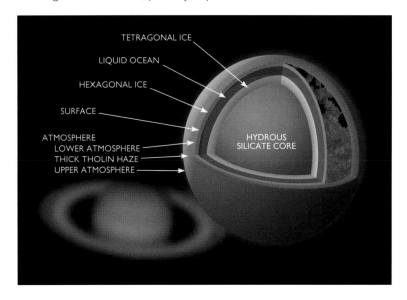

TETRAGONAL ICE

LIQUID OCEAN

HEXAGONAL ICE

SURFACE

ATMOSPHERE
LOWER ATMOSPHERE
THICK THOLIN HAZE
UPPER ATMOSPHERE

HYDROUS
SILICATE CORE

*RIGHT* **Evidence from the Cassini mission suggests there is an ocean inside Titan.** *(Data from NASA/JPL-Caltech/graphic adapted by W. D. Woods from Wikipedia/kelvinsong)*

organic compounds. The search for liquid was rewarded in 2006 when the ground track of a flyby crossed the northern region that had been in seasonal darkness since before Cassini arrived in the system, and the radar saw several well-defined patches up to 100km across whose radar reflections implied they were either a very smooth surface or a radar-absorbing substance, and liquid hydrocarbon would be both. Channels which led into or out of these patches were consistent with liquid flowing across the surface. Later studies also suggested that liquid could flow in the subsurface. The localisation of methane and ethane at the winter pole implied a seasonal cycle in which volatiles migrated from pole to pole during a local Saturnian year of 29.5 terrestrial years.

It has been speculated that anoxic conditions in the thick gloop in the lakes of hydrocarbons may have prompted the reactions necessary for the development of primitive life. Titan could therefore be a potentially life-bearing world that is now in deep freeze. It is possible it may 'bloom' in 6 billion years or so, when the Sun's helium core collapses and the sudden release of gravitational energy inflates the outer layers to produce a 'red giant' whose radius will consume the inner planets, including Earth. As the temperature on Titan rises, advanced life might emerge from its lakes to inherit the solar system.

**Future Saturn missions**
The Cassini mission was concluded in 2017 by having the spacecraft dive into the atmosphere of Saturn, to eliminate the possibility of its impacting either Titan or Enceladus and potentially contaminating those pristine environments.

Even as Cassini was active, a large number of mission concepts were devised for further studies of the Saturnian system, some of which would investigate both Enceladus and Titan while others would focus on one or the other.

At the time of writing this book in the summer of 2018, none has been funded but one is under active review. Called Dragonfly, it envisages flying a drone-like rotorcraft to study the prebiotic chemistry and habitability at numerous sites on Titan. It is in competition with several other projects to other destinations and the selection will be in the spring of 2019. In addition, NASA has funded preliminary work for a project called Enceladus Life Signatures and Habitability. This work is to assess cost-effective methods for limiting spacecraft contamination, a key factor in assessing the inclusion of life detection instruments on cost-capped missions.

**Farther out**
It would appear that any icy body in the outer solar system whose interior has sufficient heat is likely to have a layer of liquid water in its interior. The temperature at which ice melts is significantly reduced by small amounts of salts and/or methane and ammonia, the latter being abundant in the outer regions of the solar system.

An analysis in 2006 on the basis of an equilibrium between the rate that heat is produced in a rocky core and the heat that is lost through an ice shell led Hauke Hussmann, Frank Sohl and Tilman Spohn at the Institute of Planetary Research of the German Aerospace Center in Berlin to conclude that subsurface oceans were possible on Rhea (a moon of Saturn), Titania and Oberon (moons of Uranus), and Triton (a moon of Neptune), in addition to Pluto and similarly sized objects in the Kuiper belt. Even if there was only a small amount of ammonia available to serve as antifreeze, subsurface oceans could exist in such bodies.

When the New Horizons spacecraft made its Pluto flyby in 2015, the evidence of geological activity came as a considerable surprise. The bulk density implied it was 70% rock. Radiogenic heating had evidently differentiated the interior into a dense rocky core with a mantle of ice. The core was estimated to be roughly 70% of the diameter. If such heating is ongoing, there could be a subsurface ocean of liquid water 100 to 180km thick. The landforms on the large basin named Sputnik Planitia might mark where liquid water once welled up on to the surface through fractures opened by a major impact.

**The chemicals of life**
In considering possible topics for his doctorate at the University of Chicago in 1951, Stanley

Lloyd Miller attended a seminar by Harold Clayton Urey about how the synthesis of organic molecules could have occurred in conditions on the early Earth. With Urey's support, Miller performed a now famous experiment in which a flask containing a 'primordial soup' of inorganic chemicals in an appropriately 'reducing' atmosphere was irradiated with electrical sparks to simulate lightning. After running the experiment for a

**ABOVE** After decades of speculation as to the nature of Pluto, in 2015 it was revealed in all its glory by the New Horizons mission. *(NASA/JHU-APL/SwRI/Z. L. Doyle)*

**BELOW LEFT** Harold C. Urey in 1934, when he won the Nobel Prize for Chemistry. *(Nobel Foundation)*

**BELOW** Stanley Lloyd Miller with the experiment which he conducted with Harold Urey on whether the precursors for life could have formed spontaneously from a 'primordial soup' of chemicals. *(University of Chicago)*

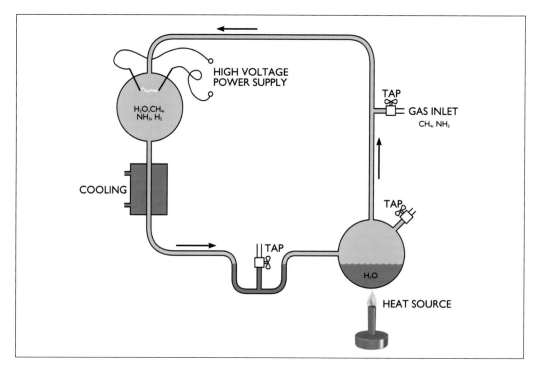

week, Miller detected the formation of amino acids. Further experiments established that fairly simple chemical processes could synthesise a wide range of organic compounds from inorganic inputs.

Because the brown, sticky substance produced by the experiment was difficult to characterise, Carl Sagan coined the term 'tholins'. These aren't a single specific compound, they are a multitude of organic compounds that are formed by solar ultraviolet irradiation and/or cosmic rays from simple carbon compounds such as carbon dioxide, methane and ethane, sometimes in combination with molecular nitrogen.

Tholins have not been produced naturally on Earth since free oxygen began to accumulate in the atmosphere several billion years ago, but they are abundant in the outer solar system on the surfaces of icy bodies and in the form of aerosols in atmospheres.

Their significance to this chapter is that in the presence of water, tholins could enable prebiotic chemistry, which has clear implications for the origin of life on Earth and possibly elsewhere in the solar system. The most common elements in terrestrial living organisms are carbon, hydrogen, oxygen and nitrogen, which are jointly known as CHON. Amino acids link up to create proteins, enzymes, lipids,

polysaccharides and nucleic acids; the building blocks and information-carrying molecules of living cells.

Tholins have been found on comets, and as comets would have rained down on the early Earth perhaps they brought not only large amounts of water but also the chemicals needed to jump-start the development of life.

Indeed, if tholins are a major constituent of the interstellar medium, we might infer that life must be ubiquitous.

In *Worlds in the Making: The Evolution of the Universe*, a book published in 1908, the Swedish chemist Svante August Arrhenius proposed that life was transported from planet to planet. He noted that spores enabled many plants, algae, fungi and protozoa to survive in unfavourable conditions for extended periods of time, and suggested that if spores were to rise to the top of an atmosphere they might leak away into space, where the radiation pressure of the host star would propel them. On arriving at another planet, a spore would come to life. This became known as the 'panspermia' hypothesis. In principle, spores could travel through interstellar space.

If not spores, then Miller's experiment suggested that at the very least, the prebiotic organic building blocks of life could be transported through space. Indeed, the raw materials of life may well have been present in the nebula from which the solar system was formed.

To paraphrase Neil Armstrong's words upon stepping on to the lunar surface, there was a "giant leap" between prebiotic compounds and the emergence of life, but in another sense it was only a "small step". Perhaps that step was taken long before the solar system formed.

Extremophiles locked into material ejected into space by an asteroid striking a planet might spend a very long time in a dormant state prior to awakening upon arrival in a different star system, perhaps on a planet or perhaps in the material of a protoplanetary disc.

So there are several scenarios. If the chemicals of life are ubiquitous, perhaps life will arise independently where conditions are conducive, giving rise to a wide variety of forms. If spores are transported between star systems, a specific kind of life will have the opportunity to establish itself in many systems.

In 1986, astronomers Fred Hoyle and Chandra Wickramasinghe expanded on panspermia by claiming that extra-terrestrial life forms entering our atmosphere are responsible for the occurrence of new diseases and the outbreak of large-scale epidemics. However, biologists have soundly rejected this idea.

## On reflection

At one time it seemed eminently reasonable that Venus, as a 'twin' of Earth, would host life. But given recent discoveries about the planet, the prospects for life as we know it having developed there have nose-dived.

Although it was once speculated that Mars might host an ancient race which was struggling to survive as their world transformed into an arid desert, this idea was discarded long before the dawn of the Space Age. At that time, it seemed that the seasonal changes of the dark areas might be vegetation. But that prospect was rejected after discovering the atmosphere to be very rarefied. Then a succession of flyby probes revealed an ancient cratered surface. In the public mind, the moniker changed from the 'Red Planet' to the 'Dead Planet'. When landers tested the soil

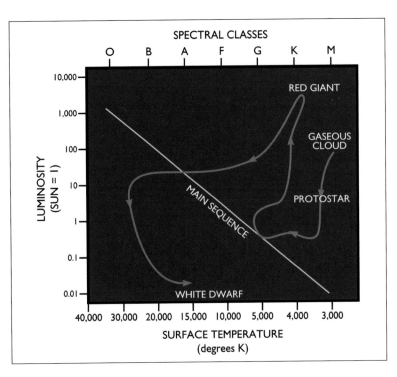

for microbes the results were ambiguous, although widely regarded as negative. We now know that conditions were more conducive to the development of life in ancient times. We may soon be able to show that life did evolve, then went extinct when surface conditions changed. Orbital analysis of trace gases might prove the planet still hosts microbial life in protected niche environments.

As regards the moons of Jupiter and Saturn, the existence of oceans under icy shells came as a surprise. Where there is a silicate ocean floor there might be extremophiles at hydrothermal vents, but it will be difficult to establish this. Perhaps the best candidate would be Enceladus because we know that there is a 'hot spot' and a mission would be able to sample the material spewing out of the fractures.

Titan is intriguing. It may have produced the organic precursors for life and is holding them in deep freeze until the Sun becomes a 'red giant' and the warmed-up moon becomes a veritable paradise. At that time too, the shells of Europa and Enceladus might melt to expose their oceans.

In the long term, therefore, as the Sun evolves away from the main sequence and renders the inner solar system uninhabitable, the locus of life may transfer to the outer realms.

**ABOVE** On the Hertzsprung-Russell diagram the evolution of a star like the Sun ends with a white dwarf. *(Data from Robert L. Carneiro/ graphic W. D. Woods)*

## Chapter Eight

# Exoplanets

After many frustrations, astronomers have made tremendous progress in finding planets orbiting other stars. The configuration of our own solar system has turned out to be far from typical. The variety observed is seriously challenging theorists. A major objective is to find planets resembling Earth in the habitable zones of their stars and then attempt to detect signals from possible inhabitants.

**OPPOSITE** The focal plane array for the Kepler Space Telescope. The wide-field camera consists of 42 CCDs, each 2.8 × 5.0cm for 1024 × 2200 pixels. *(NASA/Ball Aerospace)*

ABOVE **Bruce Campbell and Gordon Walker.**

BELOW **The dome of the Canada-France-Hawaii Telescope.** *(Sean Goebel)*

BOTTOM **The 3.6m Canada-France-Hawaii Telescope on Mauna Kea.** *(CFHT Corporation)*

# Wobbling stars

When astronomers initially sought evidence of planets orbiting around other stars, called extra solar planets or 'exoplanets', they tracked the proper motions of stars on the sky, year after year, for indications of 'wobbles'. The rationale was that although a star holds a planet in orbit, the planet tugs on the star. However, because a star is much more massive than a planet, the centre of rotation of such a system will likely remain inside the star.

For example, the presence of Earth makes the Sun rotate not around its own centre but around a point offset about 500km toward Earth. An alien astronomer in another star system would therefore require an instrument capable of detecting the Sun's travel around a circle 1,000km in diameter with a period of 1 year. And, of course, in the case of the Sun that signal would be mixed in with those from the other planets, particularly Jupiter. Even though the mass of Jupiter exceeds that of all the other solar system bodies combined, it is sufficient only to shift the centre of rotation of the solar system to a point just beyond the visible surface of the Sun.

As experience demonstrated, despite numerous false claims, this application of astrometry was unable to detect planets because the magnitude of the wobble of the star is lost in the uncertainties of the raw measurement of its position on the sky, even in the case of a planet the size of Jupiter tugging on a star as diminutive as a red dwarf that lies relatively close to us.

The low probability of success made it difficult to obtain funding and time on telescopes, so seeking planets in this way was not a promising career path. Other search techniques were required.

If the extent to which a planet would cause its star to wobble on the sky was too small to be measured using astrometric techniques, try the other great tool of astrophysics: the spectroscope. The displacement of spectral lines by the Doppler effect is a measure of the velocity along the line of sight. The problem was that the predicted shifts were so small that they, too, would be lost in the uncertainties of measuring a spectrum. A preliminary to seeking extrasolar planets was therefore to develop an instrument that had the requisite accuracy.

A conventional spectrograph drew the comparison spectrum from a lamp in the instrument. The emission lines appeared on either side of the spectrum of the target as references for measuring wavelengths. For most astronomical work this was satisfactory, but for tasks that required a sensitivity of only a few metres per second it was inadequate.

In 1979 Bruce Campbell and Gordon Walker of the University of British Columbia in Canada built a configuration in which starlight was passed through an 'absorption cell' placed between the telescope and the spectrograph so that the light for the comparison spectrum followed precisely the same route through the instrument as that for the starlight. They used hydrogen fluoride in the absorption cell, even though it was highly corrosive (indeed, if it were to escape it would be lethal) and had to be heated to 100°C. At that time, the best available resolution with other apparatus was about ±1km/sec. Theirs was capable of ±15m/sec. They installed it on the 3.6m Canada-France-Hawaii Telescope in Hawaii.

Over the next six years they and Stephenson Yang used it to routinely monitor the radial velocities of 29 stars. In 1988 they published preliminary results for 16 of those stars, several of which showed significant variations.

Of particular interest was Gamma Cephei, some 45 light years away from us. Over time its radial velocity changed by hundreds of metres per second in a way that was clearly not a perturbation from a planet. It signified a spectroscopic binary star system. The main star, a K1 orange subgiant, became Gamma Cephei A, and the companion red dwarf became Gamma Cephei B. What really attracted the team's interest was a fluctuation of ±25m/sec on the velocity of the primary that hinted at the presence of a planet with a mass several times that of Jupiter in an orbit of about 2.7 years. But the case was not solid, and in 1992 they decided there wasn't a planet.

However, in 2003 Campbell and Walker and a large team of colleagues, with a lot more data from the 2.7m telescope of the McDonald Observatory in Texas, subtracted the 57-year

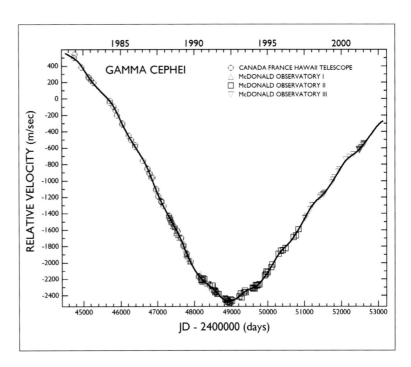

**ABOVE** The long-term radial velocity variation of the spectroscopic binary star system Gamma Cephei, annotated with the times at which data was collected for the planet that orbits the primary star.
*(McDonald Observatory press release/graphic W. D. Woods)*

**BELOW** The radial velocity curve for the planet that orbits the primary in the Gamma Cephei binary star system as obtained by the 3.6m Canada-France-Hawaii Telescope and the 2.7m Harlan J. Smith Telescope at McDonald Observatory. *(McDonald Observatory press release/graphic W. D. Woods)*

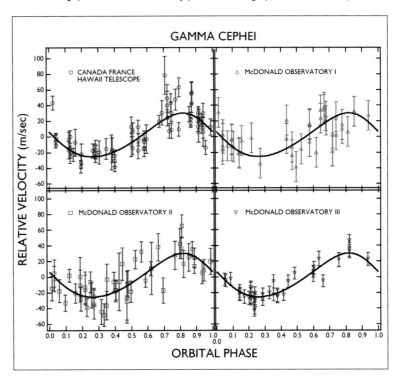

period of the binary system to isolate the rapid fluctuation. Gamma Cephei A did indeed have a planet. It had a mass at least 1.7 times that of Jupiter and was 2.13 astronomical units (au, the radius of Earth's orbit around the Sun) from its star in an orbit with a period of 2.48 years.

They had been on the right track all along, but the state of the art hadn't been sufficient to allow them to make the first unambiguous discovery of an exoplanet. When that came, it was a complete surprise.

## Pulsar planets!

As a consequence of the axial rotation of a neutron star, a radio pulsar emits brief bursts of energy at regular intervals, rather like the ticking of a clock.

When Aleksander Wolszczan used the 1,000ft Arecibo radio telescope on the island of Puerto Rico to study the pulsar PSR B1257+12 (the numbers derive from its right ascension and declination position in the sky), located about 1,000 light years away and pulsing every 6.2 milliseconds, he noticed an unusual drift in its period. Wolszczan alerted Dale Frail at the Very Large Array radio telescope in New Mexico, who confirmed this drift.

Although it was apparent that something was perturbing the neutron star, it required several

months of data to determine that it possessed two planets. They were in near-circular orbits at 0.36 and 0.46au with periods of 66.5 and 98.2 days, respectively. Their masses turned out to be 4.3 and 3.9 times that of Earth, so they were classified as 'super Earths'. This discovery was published in early 1992. Two years later, Wolszczan announced the existence of a third planet at 0.19au with a period of 25.3 days that was evidently much smaller than the others.

The existence of planets orbiting a pulsar came as a great surprise. A neutron star is the collapsed core of a star that underwent a supernova. Perhaps the blast stripped the envelopes off gas giants, leaving their cores exposed as solid planets? Alternatively, the planets could have accreted from the material ejected into space by the explosion?

**BELOW Didier Queloz (left) and Michel Mayor.** *(ESO/L. Weinstein/Ciel et Espace Photos)*

## A 'hot Jupiter'

If a spectrometer is installed on the rear of the telescope, it will suffer stress as the telescope changes targets and even while it is following a single target across the sky. Also, the internal components will adjust to environmental changes. The optical disturbances within the instrument do not normally matter, but in seeking exceedingly fine radial velocity measurements the displacements can be ruinous.

The ELODIE spectrograph was developed by André Baranne at the Marseille Observatory in France. It was an echelle type with a diffraction grating optimised for high dispersion (spacing) of spectral features on the detector, to make it easier to differentiate them. It was conventional in that a thorium-argon lamp was used to produce the comparison spectrum, but for improved stability it was installed in an environmentally controlled room and fibre optic cables fed the light from both the telescope and the lamp. With sophisticated measuring methods it was capable of ±13m/sec.

In 1993 ELODIE was installed on the 1.93m

**BELOW LEFT The ELODIE spectrograph as a museum exhibit.**
*(Observatoire de Haute-Provence/CNRS)*

**BELOW The Swiss 1.2m Leonhard Euler Telescope at La Silla with the fibre optics to feed the CORALIE spectrograph.** *(ESO)*

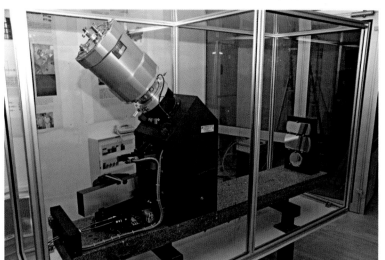

telescope at the Haute Provence Observatory in France for coverage of the northern sky, and in 1998 the similarly designed CORALIE spectrograph was installed on the 1.2m Euler telescope which the European Southern Observatory operated at La Silla in Chile, for coverage of the southern sky.

A milestone came with the announcement in October 1995 at a conference in Florence, Italy, that Michel Mayor and Didier Queloz of the University of Geneva had detected a planet orbiting the star 51 Pegasi. By convention the star itself was regarded as component 'a', so the planet was 'b'.

Queloz was working for his doctorate with Mayor as his supervisor, and his task was to analyse the ELODIE data. They started taking data in late 1994 and by March 1995 it was apparent there was a periodicity in 51 Pegasi. This was one of the brightest of a list of 100+ stars that they had chosen for study. The variation of 120m/sec was well within the capability of the instrument. What was suspicious was the cycle of only 4.2 days.

Not all radial velocity variations in starlight indicate that the star is wobbling as a result of gravitational perturbations from planets. A star is a balance between its tendency to contract under gravitation and the pressure of radiation escaping from the nuclear reactions occurring in the core. There could be pulsations which cause the envelope to inflate and contract,

varying its radius and hence imposing cyclical radial velocity changes.

Rather than publish immediately, they chose to wait for the star to become visible again in July to verify that the radial velocity oscillation was still in phase, which it was.

51 Pegasi is about 50 light years away. It has a spectral type of G5, making it similar to the Sun, which is G2. It is cooler, but its luminosity is greater because its radius is larger.

The mass of the planet is roughly half that of Jupiter. Orbiting so close to its star, at a mere 0.05au, its axial rotation is very likely to be tidally locked, with one hemisphere permanently illuminated and the other hemisphere in darkness. The heating of the upper atmosphere

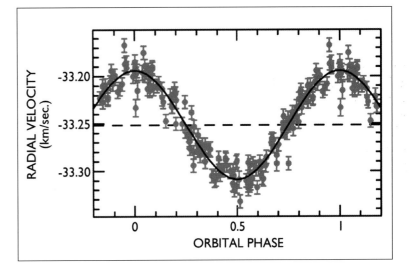

**BELOW** The radial velocity curve obtained by the ELODIE spectrograph for the star 51 Pegasi revealing a planet in an orbit with a period of 4.2 days. *(Data from M. Mayor and D. Queloz, Nature, #6555, 1995/ graphic W. D. Woods)*

ABOVE An artist's impression of the planet that was discovered by Michel Mayor and Didier Queloz orbiting the star 51 Pegasi. *(ESO/ Martin Kornmesser/ Nick Risinger)*

BELOW Artwork of a brown dwarf 'failed star'. *(NASA/JPL-Caltech)*

BELOW RIGHT Geoffrey Marcy. *(University of California, Berkeley/NASA/ Media Telecon)*

on the side facing the star has inflated the radius of the planet to a value greater than that of Jupiter. As a result, its bulk density is very low. This was the first example of a class of exoplanets appropriately named 'hot Jupiters'.

In fact, the radial velocity method favoured detecting large planets that orbit very close to stars, because their perturbations are greater. Furthermore, because the inclination of the orbital plane relative to that of the sky was not known, the 'roughly half' estimate of the mass was a minimum value. If the plane of the orbit is edge-on to us, then the inclination is 90° and we measure the actual mass. At 0°, the orbit is in the plane of the sky and we cannot see any radial velocity at all. At any other angle, the observed mass is the actual mass scaled by the sine function of the inclination angle.

The irony was that Mayor and Queloz were

not seeking planets at all; they were interested in objects with masses too large to be planets and too small to be stars. The smallest possible red dwarf is about 8% of a solar mass, which is about 80 times the mass of Jupiter. The core of a gas giant with less mass will not initiate nuclear fusion, so it will not shine in the visible spectrum. However, heat from the ongoing gravitational collapse will cause the body to radiate in the infrared. The lower transition to a planet occurs at a mass of about 15 times that of Jupiter. The intermediate objects were known as 'brown dwarfs' and astronomers were eager to find them. Since isolated ones would be difficult to detect, Mayor and Queloz decided to try to find them orbiting around normal stars. It was simple good luck that their first discovery turned out to be a planet.

Geoffrey Marcy and Paul Butler at the

University of California at Berkeley *were* seeking planets. They had an absorption cell of molecular iodine that could achieve ±3m/sec installed on a 24in auxiliary telescope at the Lick Observatory in California. Immediately on hearing the news of 51 Pegasi they gave it a look, and within a week they had sufficient data to confirm the existence of the planet.

For some time, Marcy and Butler had been measuring the radial velocities of nearby solar-type stars that were not variables. There were 120 stars on their list, but 51 Pegasi wasn't one of them. Switching priority to analysing their backlog of data, in January 1996 they were able to announce a discovery of their own. The star 70 Virginis, which is slightly farther away than 51 Pegasi, has a planet with at least 7.5 times the mass of Jupiter. The radial velocity signal was strong, with very little scatter in the data. With a period of 116 days, the planet is farther out from its parent than that of 51 Pegasi and therefore isn't so hot. However, the orbit is elliptical and the energy it receives varies considerably.

Once the radial velocity method was sufficiently refined, it made discoveries at a rapid pace. For the first time, astronomers had real data!

# The transit method

The only planets in the solar system that terrestrial astronomers can observe transiting the disc of the Sun are those whose orbits are smaller than that of Earth, namely Mercury and Venus. However, we don't see a transit every time because the planes of the orbits are not perfectly aligned.

The orbit of Venus is tilted 3.4° to that of Earth and the solar disc spans only 0.5°, so it can readily pass above or below the solar disc and be unobservable in the glare. When Venus does make a transit, we see it as a small, dark circular disc against the bright solar photosphere. During the transit, the overall light from the Sun traces a 'light curve' that drops as the planet encroaches on the disc, holds an essentially flat 'minimum' level, and finally rises to its previous level as the planet leaves the disc.

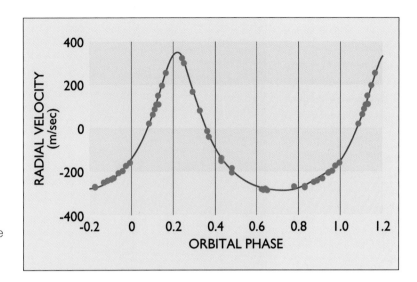

**ABOVE The radial velocity curve obtained by Geoffrey Marcy and Paul Butler for the star 70 Virginis indicating a planet in a somewhat elliptical orbit with a period of 116 days.** *(Data from exoplanets. org/graphic W. D. Woods)*

## O-B-A-F-G-K-M

The scheme for classifying the spectral types of stars derives from the efforts in the late 19th century and early 20th century to understand how stars originated and evolved. The initial alphabetic nomenclature followed what was presumed to be the evolutionary sequence. As our understanding of stars improved, the letters were retained and their ordering revised. We now know the sequence O-B-A-F-G-K-M runs from hottest to coolest. As an aide-mémoire, when astronomers see this sequence they tend to mentally recite 'Oh be a fine girl kiss me'.

Thus O and B stars are young, massive, hot, and extremely luminous. Stars of that profligacy don't last long, and are candidates to end their lives in supernova explosions. Although less massive, type A stars are still somewhat larger than the Sun and they radiate a harsh blue-white light. It is reasonable for the purposes of seeking extrasolar planets to regard types F, G, and K as broadly solar-type. The Sun is of type G, and although it formed almost 5 billion years ago it is only halfway through its hydrogen-burning life. After that it will puff up into a 'red giant'. The main sequence stars of type M are red dwarfs so faint that they will almost literally last for ever.

RIGHT Venus
transiting the disc of
the Sun in June 2012
as viewed from the
International Space
Station. *(NASA)*

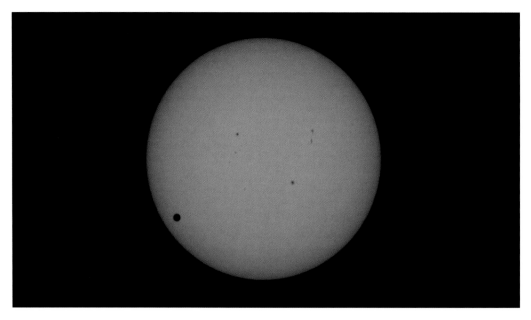

If the geometry were favourable, an alien astronomer would be able to detect Venus transiting the Sun, and from the fact that this repeated he would obtain its orbital period. In addition, the depth of the 'dip' would indicate how much light was being blocked by the planetary disc and give a measure of its size. Indeed, if the alien astronomer were patient, he would be able to observe all of the planets doing so. However, identifying the giant planets in the outer solar system would require a long-term study because it is desirable to observe a planet make several transits in order to be certain. For example, although Jupiter would yield the most prominent transit, its orbital period is 12 years and Saturn is almost 30 years.

We can use this technique to detect planets orbiting other stars, so long as the orbit is edge-on to the line of sight. Of course, we can't actually resolve the disc of a typical star, but we can detect the characteristic way that the intervening planet dims its light.

The first exoplanet to be observed in transit was orbiting HD 209458, a star in the constellation Pegasus about 150 light years away. With the spectral type G0, it is almost identical to the Sun.

Within days of the planet being discovered on 5 November 1999 by Geoffrey Marcy and Paul Butler using the radial velocity method and spectra obtained by the Keck Observatory in Hawaii, their colleague Gregory W. Henry of Tennessee State University caught it entering a transit of the star using a 0.8m automated

BELOW A light curve
of Venus transiting
the Sun in June 2012
measured by NASA's
Acrimsat. *(Data from
R. C. Willson/graphic
W. D. Woods)*

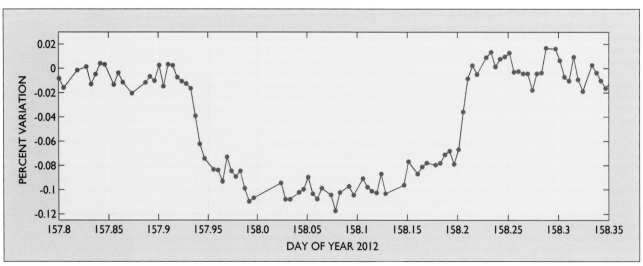

## HABITABLE ZONE

There is a volume of space around a star where conditions are 'neither too hot nor too cold but just right' for liquid water to exist on the surface of a planet with a mass comparable to Earth. This is conventionally known as the habitable zone but it is often referred to as the Goldilocks zone in reference to the children's fairy tale of 'Goldilocks and the Three Bears'.

Obviously, the location and width of the zone depends on the host star. For a G-type star, the zone will be similar to that of the Sun, whereas for an F-type star hotter than the Sun, the zone will be farther out. For a K-type star cooler than the Sun it will be closer in, and for an M-type red dwarf with a luminosity that is a tiny fraction of the Sun the zone will be very close in.

Planets in the habitable zone of a red dwarf are likely to be tidally locked, so even an otherwise very nice planet that orbits within the zone might be rendered uninhabitable if it maintains one hemisphere facing the star. But if such a planet is like Jupiter, with a family of satellites, a large moon might be habitable.

For an 'exo-Earth', we would prefer a planet to have a comparable mass to yield a similar gravitational field, and also a similar radius so that its density is comparable. Density is important because the strength of gravity on the surface of a planet depends on the radius. If a planet with a given mass is extremely dense and hence has a small radius, the surface gravity will be much stronger than for a planet with the same mass and a larger radius. We would also want a reasonable average temperature. This would vary dramatically if the orbit were significantly elliptical, perhaps so elliptical that only a fraction of the orbit was in the habitable zone of the star. And of course, it would require an appropriate atmosphere with a hydrological cycle and liquid water on the surface. In addition, the planet must rotate on its axis for a reasonable day/night cycle and the axis must be sufficiently upright to eliminate extreme seasons.

As an example to illustrate that the issue of habitability involves more than just whether the orbit of a planet lies in the zone for the host star, note that neither Venus nor Mars, respectively near the inner and outer fringes of the habitable zone of the Sun, fulfils all of these requirements.

Although our concept of life is based on a single example, it seems reasonable to suppose that microbes that developed independently in different star systems would have a lot in common. However, we shouldn't presume that this will be so for intelligent life.

Life developed very early in Earth's history, but for several billion years the biosphere consisted only of individual cells, sometimes organised into colonies. It was only about a billion years ago that primitive plants developed, and half a billion years later that animal life proliferated in what is known as the 'Cambrian explosion'. On that timescale, intelligent life arose 'in the blink of an eye'. The likelihood of intelligence developing also depends on whether the energy output of the host star remains constant. After a slow brightening early on, the Sun has evidently been stable for billions of years.

In 2013, Geoffrey Marcy suggested that perhaps as many as 20% of solar-type stars could host planets comparable to Earth with liquid water on the surface and, if the necessary chemical ingredients were present, some form of life.

**BELOW** The location and size of a habitable zone depends on the luminosity of the star. *(PHL/University of Puerto Rico at Arecibo)*

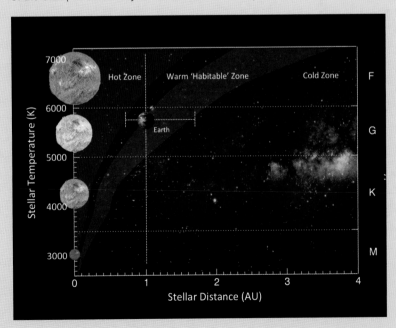

## A SPECTROGRAPH FOR SEEKING EXO-EARTHS

The High Accuracy Radial velocity Planetary Search (HARPS) spectrograph was the successor to both ELODIE and CORALIE. The principal investigator was Michel Mayor. Development work began in 1998 and when it was commissioned on the 3.6m European Southern Observatory telescope at La Silla in Chile in 2003 it achieved an accuracy of 1m/sec.

HARPS was itself succeeded by the third-generation Echelle Spectrograph for Rocky

**LEFT** The HARPS spectrograph. *(ESO)*

**ABOVE** The dome of the 3.6m telescope at La Silla in the Atacama Desert of Chile. *(ESO)*

**BELOW** The 3.6m telescope of the European Southern Observatory at La Silla on which the HARPS spectrometer is operated. *(ESO/C. Madsen)*

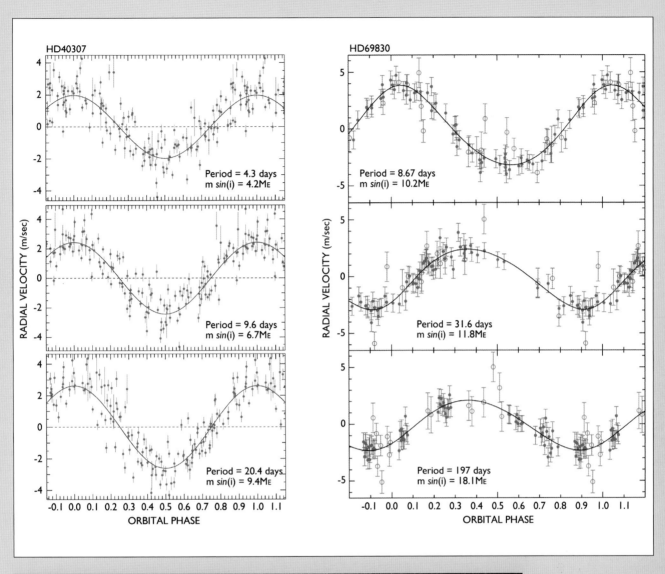

HD40307

Period = 4.3 days
m sin(i) = 4.2Mᴇ

Period = 9.6 days
m sin(i) = 6.7Mᴇ

Period = 20.4 days
m sin(i) = 9.4Mᴇ

HD69830

Period = 8.67 days
m sin(i) = 10.2Mᴇ

Period = 31.6 days
m sin(i) = 11.8Mᴇ

Period = 197 days
m sin(i) = 18.1Mᴇ

RADIAL VELOCITY (m/sec)

ORBITAL PHASE

**ABOVE** Examples of radial velocity curves obtained by HARPS. *(Data from ESO/graphic W. D. Woods)*

**LEFT** The four main domes of the Very Large Telescope of the European Southern Observatory at Paranal in the Atacama Desert of Chile. In the background are several smaller auxiliary facilities. *(ESO/Miguel Claro)*

Exoplanets and Stable Spectroscopic Observations (ESPRESSO), which was installed at Cerro Paranal in Chile in 2017, where it will be able to draw light from the four 8.2m elements of the Very Large Telescope. It was designed to achieve an accuracy of better than 3cm/sec. If fine tuning improves that to 1cm/sec, it will be a factor of 100 better than HARPS.

An alien equipped with such an instrument would be able to detect the radial velocity of the Sun induced by Earth, which is 10cm/sec. We, in turn, will be able to do likewise for exo-Earths in the habitable zones around other solar-type stars, although of course it will be necessary to monitor for a year to see a single cycle of the radial velocity curve.

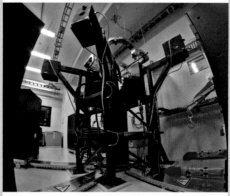

**RIGHT** The ESPRESSO spectrometer draws light from the four 8.2m units of the Very Large Telescope of the European Southern Observatory in Chile (top, *ESO/L. Calçada*). The images below show the instrument chamber, the four feeders, the control room for the November 2017 'first light', and data displays. *(INAF Trieste/Giorgio Calderone)*

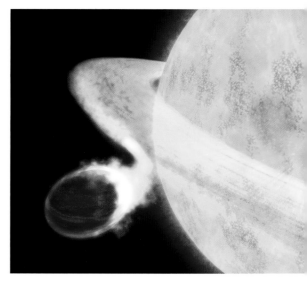

**ABOVE** Artwork of the atmosphere of the planet WASP-12b 'boiling away' into space. *(NASA/ESA/STScI/G. Bacon)*

select those that were varying in a manner characteristic of a transiting planet and they would then be investigated for radial velocity evidence of a planet.

Not surprisingly, starting in 2006 most of the exoplanets identified by WASP were 'hot Jupiters'. The 12th discovery, in 2008, was a planet orbiting a G0 star in the constellation of Auriga with a period of 1.1 days. Its mass was 1.4 times that of Jupiter, and because the radius of its orbit was a mere 0.02au it was so hot that its atmosphere had to be 'boiling away' to space. This is a runaway process because the more mass the planet loses, the less it is able to hold on to what remains of the atmosphere. As a result, in 10 million years or so it will have been reduced to its rocky core.

Once the transit method had been proven, a variety of projects were set up to seek the glory of discoveries. Some of these were strictly professional, but others included amateur astronomers. Automated systems were preferred. With all this activity, the number of 'hot Jupiters' rapidly increased. But the prize for a planet hunter was to find one with a mass similar to Earth, hopefully orbiting within the 'habitable zone' of its star.

In 2006 the Centre National d'Études Spatiales (CNES), the space agency of France, launched into a 900km polar orbit a satellite

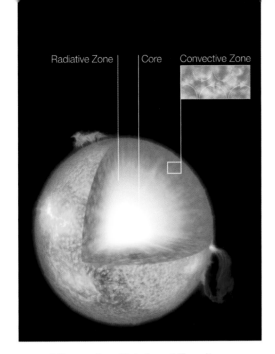

Radiative Zone | Core | Convective Zone

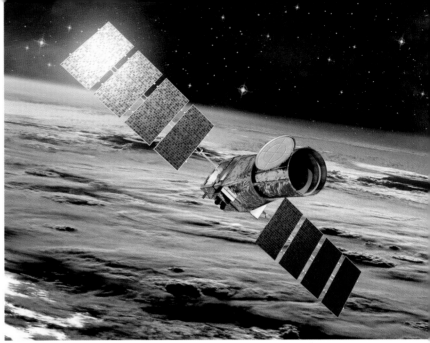

**ABOVE LEFT** In addition to seeking transiting planets, the CoRoT mission was to study the acoustic signatures of convection within solar-type stars. *(ESO)*

**ABOVE** Artwork of the CoRoT satellite. *(CNES/David Ducros)*

named Convection, Rotation et Transits planétaires (CoRoT) with the dual purpose of studying the interiors of stars by way of their surface vibrations (in the case of the Sun, this is known as helioseismology) and, if they were present, to detect transiting planets, possibly a 'super Earth' with a mass only twice that of our own planet.

CoRoT had a 0.27m telescope equipped with four 2048 × 4096 CCDs, two for the seismological investigations and two for transiting planets. It spent six months observing a field of stars on one side of the sky then, when the Sun intervened, it switched to a field on the other side of the sky. It alternated back and forth in this manner until the mission was declared finished in 2013. It operated by taking a continuous sequence of exposures, each lasting 32sec, but the image was not fully transmitted to Earth because that data flow would be excessive. The computer on the spacecraft used a mask to choose the pixels surrounding each of the designated target stars, summed all of the pixels within the mask, and then added successive exposures to attain an integration time of about 8min. It was this that was reported. A finer time resolution could be achieved for any specific case by transmitting each 32sec exposure for an observation lasting several hours. The project was supported by a similarly sized telescope in Chile operated by the Berlin Exoplanet Search.

CoRoT found its first exoplanet within months of setting to work. This was in orbit around a solar-type star some 1,500 light years away

in the constellation of Monoceros. Its second discovery was orbiting a star about 1,000 light years away. Both were 'hot Jupiters'.

The seventh discovery, in 2009, was notable for being the smallest exoplanet so far. The star,

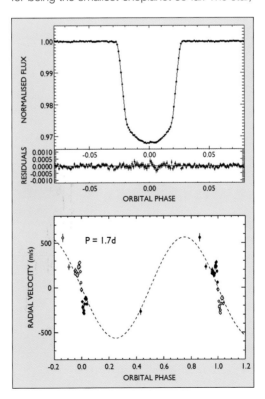

**LEFT** The transit light curve of planet CoRoT-2b and the matching radial velocity plot by Geoffrey Marcy et al. *(Data from exoplanets.org/graphic W. D. Woods)*

also in Monoceros and about 500 light years away, was of spectral type G9, making it a little cooler than the Sun. The planet was orbiting around it at a distance of only 0.017au with a period of 20.5hr. It was calculated to have a mass of about six times that of Earth with a radius some 70% greater, implying a similar bulk density. It was therefore a rocky 'super Earth'. As it would be tidally locked, the hemisphere that permanently faced the star must be several thousand degrees. The temperature of the dark hemisphere must be a mere 50K. The mass

wasn't enough for the planet to have gained a hydrogen envelope. Any original atmosphere would either have been lost to space or frozen on to the surface on the dark hemisphere. If there is an atmosphere now, it will derive from the superhot surface and it will migrate around to the dark hemisphere, to freeze out as a form of metal-rich snow.

High-performance radial velocity data from the HARPS spectrograph on the 3.6m telescope at La Silla Observatory in Chile seeking confirmation of this planet gave a surprise in the form of a second 'super Earth' whose orbit did not produce transits. It was farther out, at 0.046au, with a period of 3.7 days. Its mass was rather less well defined but could be twice that of its companion, in which case it may well have retained a primordial envelope. That would make it an example of a 'hot Uranus'.

CoRoT's ninth discovery, reported in 2010, orbited a star roughly 1,500 light years away in the constellation of Serpens. With a spectral type of G3, the star was similar to the Sun. The planet was slightly less massive than Jupiter, with a radius that was comparable. At 0.41au with a period of 95 days, it was outside the realm of the 'hot Jupiters', making it a 'temperate' giant with a primordial atmosphere.

While CoRoT sought 'super Earths' orbiting

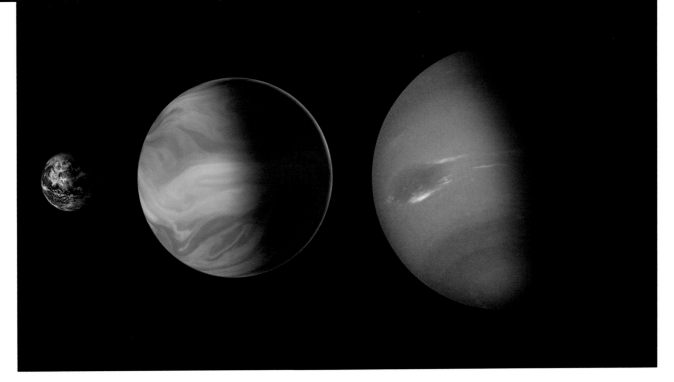

solar-type stars, astronomers on the ground were pursuing a different strategy.

The MEarth Project funded by the National Science Foundation was similar to WASP. It had two sites, one at the Whipple Observatory on Mount Hopkins in Arizona to cover the northern hemisphere and the other at the Cerro Tololo Inter-American Observatory in Chile. Each had eight 0.4m, f/9 robotic telescopes that were mounted independently rather than in a group. Each telescope had a 2048 × 2048 CCD. The plan was to monitor red dwarfs, firstly because they are far more numerous and also because a transiting planet would make a deeper light curve than the same planet would with a star like the Sun. This improved the chances of detecting smaller planets.

At the end of 2009 the project reported a 'super Earth' was transiting a star 40 light years away in the constellation of Ophiuchus. The star is designated GJ-1214 because it is in the *Catalogue of Nearby Stars* that was issued in 1979 by German astronomers Wilhelm Gliese and Hartmut Jahreiß. Its luminosity is a tiny fraction of that of the Sun, and there is a 1% variability in the near-infrared that very likely indicates the presence of 'starspots' (similar to 'sunspots', but more prominent) which also account for flare activity that generates X-rays (as is common for such stars). The planet, whose existence was confirmed by radial velocity observations, has a mass of about six times that of Earth. Although it orbits at only

0.014au with a period of 1.58 days, the dimness of the star means the surface of the planet can't exceed 200°C. It could possess surface water and a dense atmosphere, but a near-infrared study of it in transit by the Very Large Telescope on Cerro Paranal in the Atacama Desert of Chile ruled out a hydrogen envelope. It could have started out with one, only to have it stripped away by the X-rays from the flaring of the star.

At last, therefore, the transit method was beginning to reveal 'rocky' planets, even if they were still somewhat larger than Earth.

**ABOVE** Artwork comparing the planet GJ-1214b with Earth and Neptune. *(NASA/ESA/STScI/A. Feild/G. Bacon)*

**BELOW** Artwork depicting the planet GJ-1214b transiting its 'spotty' parent star. *(ESO/L. Calçada)*

# The Kepler 'fire hydrant'

The NASA Kepler Space Telescope was designed to detect transits by Earth-sized planets in the habitable zones of solar-type stars as far as 3,000 light years away.

It was launched in 2009 to orbit independently around the Sun with a period 6 days longer than an Earth year to cause it to slowly recede. This isolation was to enhance its ability to accurately point in a fixed direction throughout its operating life.

It had a Schmidt camera with a mirror 1.4m in diameter that provided a field of view of 100 square degrees. The enormous detector had a mosaic of 42 CCDs, each 1024 × 2200 pixels. The telescope was to stare continuously at a rich star field of the Milky Way midway between the stars Deneb in the constellation of Cygnus and Vega in Lyra, measuring the brightness of over 150,000 stars in the Orion arm of the galaxy every 6sec.

The plan was for Kepler to provide data for 3.5 years, which was long enough to observe three transits by exoplanets in Earth-like orbits to verify their existence. An extension to the

**ABOVE** Artwork of the Kepler Space Telescope, named in honour of the 17th century astronomer Johannes Kepler. *(NASA)*

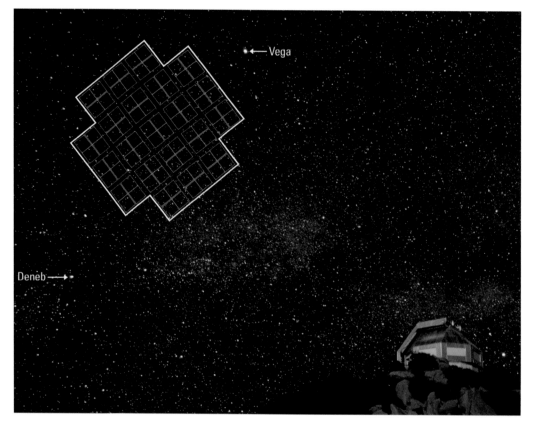

**RIGHT** The field of the camera on the Kepler Space Telescope superimposed on a ground-based image of the sky from Kitt Peak, Arizona, with the dome of the 3.5m telescope which is operated by a consortium involving the University of Wisconsin-Madison, Indiana University, Yale University, and the National Optical Astronomy Observatories. *(KPNO)*

Vega

Deneb

mission would increase the chances of seeing a single transit by a giant planet farther out, and those would be pursued by ground-based radial velocity observations.

Kepler started its observations in May 2009, and by the end of the year it had noted two previously known planets and had discovered five of its own: four 'hot Jupiters' and a 'hot Neptune'. In mid-2010 it found three planets orbiting a solar-type star around 2,000 light years away, designated Kepler-9. Two of the planets had masses similar to Saturn and were in orbits of 0.14 and 0.22au with periods of 19 and 39 days. The third was a 'super Earth' with a mass 7 times that of Earth. Orbiting at 0.03au with a period of 1.6 days, it had to be very hot.

The tenth entry in the Kepler list was orbiting a solar-type star 560 light years away. At 0.02au with a period of 0.84 days, the planet had a mass 4.6 times that of Earth and a radius 40% larger, so with a bulk density twice that of Earth it must possess a very large metal core.

The 11th discovery orbited a star resembling the Sun located 1,700 light years away. This marked a major milestone because the Kepler-11 star had six planets. The outermost, which was comparable in mass to Neptune, was at 0.46au with a period of 118 days. The other five were all in orbits between 0.09 and 0.25au with periods ranging from 10 to 47 days. Their masses ranged from 2 to 8 times that of Earth and their radii were 2 to 4 times that of Earth. Amazingly, this entire system of planets would fit comfortably inside the orbit of Venus in our system.

By 2011 the Kepler mission had identified

over 1,600 candidates. Work was underway to verify them by seeking confirmatory transits and by radial velocity observations. Of these, several dozen were in, or close to, the habitable zone of the host star. The objective remained to detect potentially habitable planets that were comparable to Earth.

At the time of writing this book (mid-2018) the system with the most number of planets is Kepler-90, a G-type star 2,500 light years away. With eight planets, the system was distinctive for its similarity to the solar system with rocky planets in close and giants farther out, although

**ABOVE Artwork of the planets of the Kepler-11 system.** *(NASA/Tim Pyle)*

**BELOW The Kepler-90 system.** *(NASA/AMES/ Wendy Stenzel)*

RIGHT The facility at the European Southern Observatory at La Silla in Chile operated by the Transiting Planets and Planetesimals Small Telescope (TRAPPIST) network. *(ESO)*

BELOW The transits of the planets in the TRAPPIST-1 system as observed by the Spitzer Space Telescope. *(Data from ESO/M. Gillon/ annotation W. D. Woods)*

BELOW RIGHT The 0.6m TRAPPIST telescope at La Silla in Chile. *(ESO)*

it occupies a volume of space that would fit within the orbit of Earth. Apart from the two outer giants, the others are either 'super Earths' or 'mini Neptunes'. One of the planets was identified in the transit data in 2017 using a machine-learning method developed by Google.

The next most populous system, with seven planets, was the first discovery by the Transiting Planets and Planetesimals Small Telescope (TRAPPIST) project that used a facility at La Silla in Chile for the southern sky and another in Morocco for the northern sky, each having a robotic 0.6m telescope. The star was an ultra-cool red dwarf with 8% the mass and 11% the

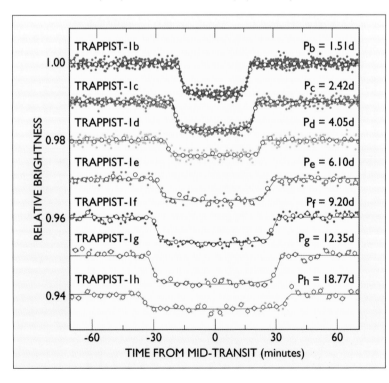

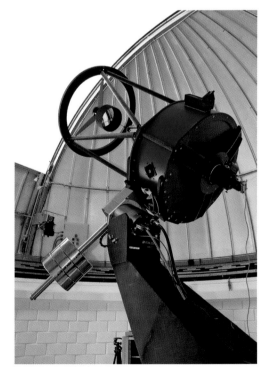

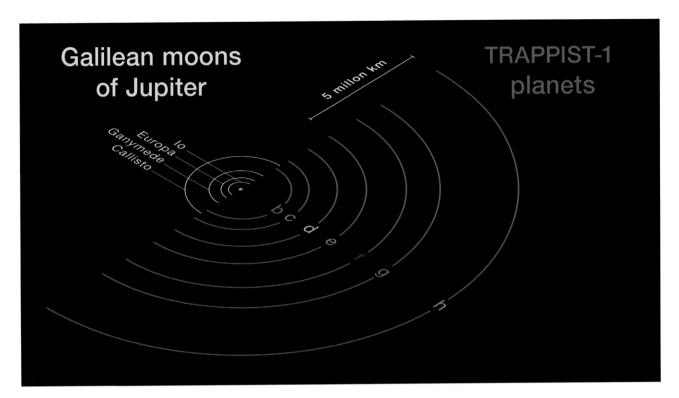

**Galilean moons of Jupiter**

Europa
Ganymede
Callisto
Io

5 million km

**TRAPPIST-1 planets**

b c d e f g h

ABOVE **The TRAPPIST-1 system.** *(ESO/O. Furtak)*

radius of the Sun, located 40 light years away in the constellation Aquarius. In 2015 a Belgian team identified three Earth-sized planets by transits, and another four were found by follow-up studies using the Spitzer Space Telescope and the Very Large Telescope in Chile. This system is so compact that it is comparable to the Jovian satellites in our system. With so many planets in such close proximity, it was possible to accurately measure their masses. Five are similar in size to Earth and two are intermediate between Mars and Earth. They are all 'temperate', and three are within the habitable zone of the star.

For so long, theorising about how a planetary system formed had been based on the single example of our own. With data for over 3,000 confirmed exoplanets, obtained using a variety of detection methods, it is apparent that our solar system is far from typical.

Within a single generation, astronomers had gone from being starved of data to being deluged with it. Indeed, the flow from the Kepler mission was likened to attempting to drink water from a fire hydrant.

Once we have theories which can explain the bewildering variety of systems, we will have taken an enormous step forward in understanding the universe.

## TRANSITING EXOPLANET SURVEY SATELLITE

Kepler continued monitoring until the summer of 2013, halting only when the spacecraft lost its ability to point with the desired precision. It was then assigned a less demanding programme that would involve seeking habitable planets around smaller, dimmer red dwarfs, working its way around the band of sky in the plane of Earth's orbit. The propellant supply enabled it to operate in this mode through into the summer of 2018.

By that time, its successor, the Transiting Exoplanet Survey Satellite (TESS), was operating in an inclined high orbit around Earth that permitted unobstructed imaging of both the northern and southern hemispheres of the sky.

During the two-year duration of its primary mission, a linear array of four 16.8 megapixel cameras that together span $24 \times 96°$ will create an all-sky survey. It is to pay particular attention to small rocky planets orbiting nearby G, K, and M-type stars. Based on Kepler statistics, it is expected to discover at least 20,000 transiting exoplanets, many of them similar to Earth and situated in habitable zones. It will thus provide prime targets for further study by the James Webb Space Telescope, as well as future large ground-based systems and space-based telescopes that will assess the prospects for their hosting life.

# Drake's equation

In September 1959 the journal *Nature* ran an article by the eminent physicists Giuseppe Cocconi and Philip Morrison with the provocative title 'Searching for Interstellar Communications'. They pointed out that radio telescopes had become sufficiently sensitive to detect transmissions that might be broadcast into space by civilisations orbiting nearby stars.

Shortly thereafter, radio astronomer Frank Drake of Cornell University used the 85ft antenna at the National Radio Astronomy Observatory at Green Bank in West Virginia for

## EXOPLANET BIOSIGNATURES

It is very difficult to directly observe a planet in orbit around a solar-type star because in the visible spectrum the planet shines only by the light that it reflects and the star outshines the planet by a factor of a billion. In the infrared, however, a star like the Sun is intrinsically less luminous and the heat radiated by a planet reduces the ratio to one in a million. This provides us with an excellent opportunity to seek evidence of life on Earth-like planets in the habitable zones of stars similar to the Sun.

Once we have the means to isolate the spectrum of such a planet from its star, a major objective will be to examine the absorption lines in the wavelength range 6 to 18μm since this contains lines of water, ozone, and carbon dioxide. A strong carbon dioxide line would show that the planet has an atmosphere. Ozone would show that the atmosphere contains oxygen. As oxygen is highly reactive, for it to exist 'out of equilibrium' in an atmosphere there must be a continual source of replenishment. Its presence would be a strong indicator of life. The oxygen which makes up 21% of the terrestrial atmosphere was created by, and is maintained by, life. And a planet with absorption lines of water would likely possess substantial bodies of open water on its surface. Finding such an atmosphere would clinch the case for life.

But first, we must find some true exo-Earths.

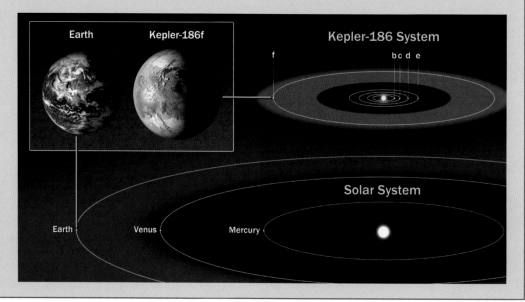

Project Ozma, which was the first attempt to seek such signals. He 'listened' to Tau Ceti and Epsilon Eridani, two stars about a dozen light years away and generally similar to the Sun. Following Cocconi and Morrison's advice, Drake slowly scanned frequencies close to the 21cm line of neutral hydrogen for six hours a day from April to July 1960 with negative results. At that time no one knew whether either star had any planets to host a technological civilisation, but recent studies have suggested that each possesses at least one planet.

In 1961, in preparing for the first meeting of the Search for Extra-Terrestrial Intelligence (SETI), Drake decided to use a mathematical expression to structure the debate. As an equation, this probabilistic rationale might not be as famous as Einstein's equivalence of

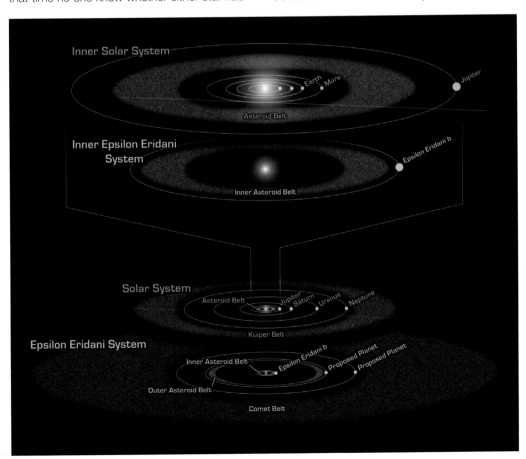

mass and energy but it has certainly made its mark.

Drake based his calculation on the following factors:

$R_*$ — the average rate at which stars are formed

$f_p$ — the fraction of such stars that have planets

$n_e$ — the average number of planets orbiting such stars that have the potential to support life

$f_l$ — the fraction of those planets that actually develop life

$f_i$ — the fraction of planets bearing life on which intelligence has evolved

$f_c$ — the fraction of civilisations that develop the means to communicate over interstellar distances

$L$ — the length of time that such civilisations emit detectable signals

And he defined the number of active, communicative extra-terrestrial civilisations in the Milky Way system, N, as the product of these factors:

$$N = R_* \times f_p \times n_e \times f_l \times f_i \times f_c \times L$$

Of course, this equation was not meant to provide a definitive answer; its role was to introduce the concepts that would influence the likelihood of our detecting such civilisations.

The difficulty with a computation of this nature is that it rests on the proper framing of the individual probabilities. In the absence of data, none of the factors raised by Drake were quantifiable at that time.

We now possess more data on the planetary systems of other stars than was available in 1960 and we know considerably more about the nature of life, but we know nothing more of extra-terrestrial intelligence. So, although we may be better able to frame the probabilities for some of Drake's factors, we still know next to nothing of the others, any one of which could skew the result decisively one way or the other.

Furthermore, it was presumed that alien civilisations would communicate by radio. It has recently been argued that they would be more likely to communicate using lasers. In effect, as our technologies evolve, we project our new capabilities on to aliens and then ponder how we might detect such activities.

In truth, the Drake equation is really just a narrow aspect of the more general Fermi paradox.

**RIGHT At the first Search for Extra-Terrestrial Intelligence (SETI) meeting in 1961 at Green Bank, West Virginia, Carl Sagan outlines the equation devised by organiser Frank Drake.**

# Fermi's paradox

After reflecting upon the apparent contradiction between the high probability of the existence of extra-terrestrial civilisations and the lack of evidence for them, the renowned physicist Enrico Fermi casually asked, "Where is everybody?" This question has become known as the Fermi paradox.

The Milky Way system contains billions of stars which are similar to the Sun. Although the universe is almost 14 billion years old, the Sun was formed when an interstellar cloud collapsed almost 5 billion years ago, meaning many of the solar-type stars are much older than the Sun. It appeared reasonable that many of these stars would possess planets similar to Earth. Furthermore, it seemed likely, on the basis of the fact that we exist, that intelligent life may have developed on some of those planets. If an alien civilisation developed a means of travelling between the stars, then even progressing at sub-light speed it could explore the galaxy in a few million years. Hence Fermi's interrogative: Where are they?

One way to resolve this paradox is to suggest that intelligent life is extremely rare. Perhaps our civilisation is the first. Perhaps

**LEFT Enrico Fermi in the mid-1940s.** *(US Department of Energy)*

there are many civilisations that stay within the confines of their own planetary system. Perhaps they are aware of one another and communicate. If so, we might one day detect them. We are on the verge of gaining the ability to explore the farthest reaches of our solar

**LEFT The Allen Telescope Array at the Hat Creek Radio Observatory in California is listening for signals from extra-terrestrial intelligence.** *(SETI Institute)*

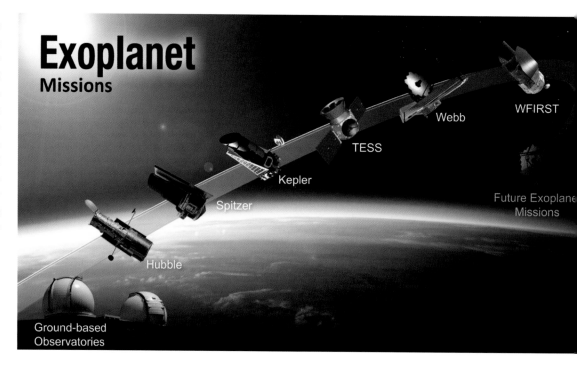

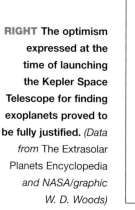

RIGHT Exoplanet telescopes. Transiting Exoplanet Survey Satellite entered service in 2018. The much-delayed launch of the James Webb Space Telescope is currently scheduled for 2021. If the Wide Field Infrared Survey Telescope is built, it is expected to enter service in the mid-2020s. *(NASA)*

**Exoplanet**
**Missions**

WFIRST

Webb

TESS

Kepler

Spitzer

Future Exoplanet Missions

Hubble

Ground-based Observatories

system, and in doing so we might happen across relics of travellers who passed through it long ago. Perhaps it is intrinsic to even galaxy-spanning civilisations that they last only a few million years, in which case when we set forth in interstellar ships we might find relics left by many generations of our predecessors.

Alternatively, of course, the galaxy could contain many civilisations which, in accordance with a 'prime directive' that mandates non-interference, are giving us a wide berth until we are deemed sufficiently mature to deal with 'first contact'. The evidence of our own history suggests that a primitive society is disrupted by contact with a more advanced one. In the context of our discovering that we

RIGHT The optimism expressed at the time of launching the Kepler Space Telescope for finding exoplanets proved to be fully justified. *(Data from* The Extrasolar Planets Encyclopedia *and NASA/graphic W. D. Woods)*

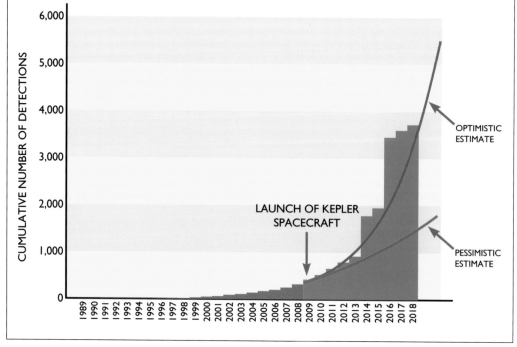

LAUNCH OF KEPLER SPACECRAFT

OPTIMISTIC ESTIMATE

PESSIMISTIC ESTIMATE

are the juveniles in a galaxy-wide society, no matter how familiar we may be with the laws of nature and how we apply them to produce technologies, it will come as a shock to realise how little we actually know. As Arthur C. Clarke observed, "Any sufficiently advanced technology is indistinguishable from magic."

As our investigations determine which stars possess planets in their habitable zones and which don't, this information can be used to target future searches for extra-terrestrial life by means of radio signals, laser beams, or some other method that we might dream up.

For example, in 2016, the Allen Telescope Array sited at the Hat Creek Radio Observatory in California targeted the TRAPPIST-1 system on behalf of the SETI Institute and scanned methodically through 10 billion radio channels in search of evidence of an alien presence…with negative results.

## So what is it all for?

The Anthropic Principle is a philosophical consideration that observations of the universe must be compatible with the conscious and sapient life that observes it. It is remarkable that just a few numbers, physical constants of nature, define the key features of the cosmos. If the value of any one of these numbers had been significantly different, this could have made it impossible for life to develop. If a 'multiverse' of parallel universes exists in which the values of these numbers are different in each case, then our very existence can be attributed to the fact that this particular universe is configured for life. Indeed, it might even be argued that the purpose of our universe was to give rise to life. Perhaps we are simply part of a 'double blind' experiment of some sort that is being conducted by beings who live in another realm!

**BELOW** Artwork of the Extremely Large Telescope (ELT) which, with its 40m diameter segmented main mirror, will be the largest optical and near-infrared telescope when it becomes available in 2024 at the European Southern Observatory at Cerro Armazones in the Atacama Desert of Chile. *(ESO)*

# Postscript

For thousands of years we thought the universe consisted of Earth, its Moon, the Sun, and specks of light which wandered across the celestial realm against the background of fixed lights. We now know it is much larger and much richer in its constituents.

Our scientific investigation began with Isaac Newton, whose laws of physics, in particular, his law of gravitation, explained so much that had been mysterious. With the 20th century theories of relativity and quantum mechanics, we were able to probe back to the merest fraction of a second after the act of creation, called the Big Bang, almost 14 billion years ago. Along the way we realised there was a brief period in which the radius of the universe underwent an exponential increase by doubling 100 times. Then expansion was slowed by the gravitational attraction of material in the universe, at a rate given by the Hubble Constant. We also realised that the matter that we can see by the electromagnetic radiation it emits is less than 5% of the total; the identity of the rest, dubbed 'dark matter', is currently a mystery. Worse, matter itself is a minor constituent of the universe! The dominant agent manifests itself as a force that is overwhelming self-gravitation and causing the rate of expansion to accelerate. Although we have labelled this 'dark energy', we don't actually know what it is.

And closer to home, there is the matter of ourselves. As yet, Earth is the only world on which we know there to be life. We are now making strenuous efforts to determine whether it exists elsewhere, including in oceans beneath the icy crusts of bodies in the outer solar system.

In recent years, we have gone from wondering whether any other stars have systems of planets to discovering that they are common, and the search is on for a star which has an Earth-like planet with an atmosphere whose composition is 'out of equilibrium' in a manner indicative of life.

If it turns out that life has originated independently on different bodies in our own star system, and also in other systems, then we might reasonably infer that it is likely to be ubiquitous in the universe. And that raises the question of whether there are alien civilisations out there.

**OPPOSITE An artist's impression of an ultra-cool red dwarf some 40 light years away that has five planets that are similar in size to Earth, plus two that are slightly smaller. Three planets lie within the habitable zone of the star. It is possible that such systems might host some form of life.** *(ESO/Martin Kornmesser)*

# Further reading

*(in chronological order)*

J. B. Sidgwick, *William Herschel: Explorer of the Heavens*, Faber and Faber, 1953

M. A. Hoskins, *William Herschel and the Construction of the Heavens*, Oldbourne, 1963

William Bixby, *The Universe of Galileo and Newton*, Cassell, 1966

Patrick Moore, *The Astronomy of Birr Castle*, Mitchell Beazley, 1971

Fred Hoyle and Chandra Wickramasinghe, *Lifecloud: The Origin of Life in the Universe*, Dent, 1978

Robert W. Smith, *The Expanding Universe: Astronomy's 'Great Debate' 1900-1931*, Cambridge University Press, 1982

Gale E. Christianson, *In the Presence of the Creator: Isaac Newton and His Times*, Macmillan, 1985

Frank Drake and Dava Sobel, *Is Anyone Out There? The Scientific Search for Extraterrestrial Life*, Delacorte Press, 1992

Stephen Hawking, *Black Holes and Baby Universes*, Bantam, 1993

George Smoot and Keay Davidson, *Wrinkles in Time: Imprint of Creation*, Little, Brown, 1993

Paul Davies, *Are We Alone? Philosophical Implications of the Discovery of Extraterrestrial Life*, Basic Books, 1995

John C. Mather and John Boslough, *The Very First Light: The True Inside Story of the Scientific Journey Back to the Dawn of the Universe*, Basic Books, 1996

Stephen Hawking, *The Illustrated A Brief History of Time (Updated and Expanded Edition)*, BCA, 1996

Gale E. Christianson, *Edwin Hubble: Mariner of the Nebulae*, Institute of Physics, 1997

David Grinspoon, *Venus Revealed: A New Look Below the Clouds of our Mysterious Twin Planet*, Addison-Wesley, 1997

Paul Davies, *The Fifth Miracle*, Penguin, 1998 (reissued as *The Origin of Life* in 2003)

Stephen Webb, *Measuring the Universe: The Cosmological Distance Ladder*, Springer-Praxis, 1999

Martin Rees, *Just Six Numbers*, Weidenfeld & Nicolson, 1999

Freeman Dyson, *Origins of Life (2nd edition)*, Cambridge University Press, 1999

Alan W. Hirshfeld, *Parallax: The Race to Measure the Cosmos*, W. H. Freeman, 2001

Ralph A. Alpher and Robert Herman, *Genesis of the Big Bang*, Oxford University Press, 2001

Michael Hanlon, *The Worlds of Galileo: The Inside Story of NASA's Mission to Jupiter*, St. Martin's Press, 2001

Michael D. Lemonick, *Echo of the Big Bang*, Princeton University Press, 2003

Fulvio Melia, *The Black Hole at the Center of Our Galaxy*, Princeton University Press, 2003

Barrie W. Jones, *Life in the Solar System and Beyond*, Springer-Praxis, 2004

George Johnson, *Miss Leavitt's Stars: The Untold Story of the Woman Who Discovered How to Measure the Universe*, W. W. Norton, 2005

Manjit Kumar, *Quantum: Einstein, Bohr and the Great Debate about the Nature of Reality*, Icon Books, 2008

Ralph Lorenz and Jacqueline Mitton, *Titan Unveiled: Saturn's Mysterious Moon Explored*, Princeton University Press, 2008

Richard Greenberg, *Unmasking Europa: The Search for Life on Jupiter's Ocean Moon*, Copernicus, 2008

Alan Boss, *The Crowded Universe: The Search for Living Planets*, Basic Books, 2009

Brian Cox and Andrew Cohen, *Wonders of the Solar System*, Collins, 2010

John Farrell, *The Day Without Yesterday: Lemaître, Einstein, and the Birth of Modern Cosmology*, Basic Books, 2010

Marcus Chown, *Afterglow of Creation: Decoding the Message from the Beginning of Time (Updated Edition)*, Faber and Faber, 2010

Louis Neal Irwin and Dirk Schulze-Makuch, *Cosmic Biology: How Life Could Evolve on Other Worlds*, Springer-Praxis, 2011

David Schultz, *The Andromeda Galaxy and the Rise of Modern Astronomy*, Springer, 2012

Marcia Bartusiak, *Black Holes: How an Idea Abandoned by Newtonians, Hated by Einstein, and Gambled on by Hawking Became Loved*, Yale University Press, 2015

Stephen Webb, *If the Universe is Teeming with Aliens – Where is Everybody? Seventy-Five Solutions to Fermi's Paradox and the Problem of Extraterrestrial Life (2nd edition)*, Springer, 2015

Stuart Clark, *The Search for Earth's Twin*, Quercus, 2016

Dava Sobel, *The Glass Universe: The Hidden History of Women Who Took the Measure of the Stars*, Fourth Estate, 2016

Jim Al-Khalili (ed), *Aliens, Science Asks: Is There Anyone Out There?*, Profile Books, 2016

Elizabeth Tasker, *The Planet Factory: Exoplanets and the Search for a Second Earth*, Bloomsbury Sigma, 2017

David M. Harland, *Mars Manual*, Haynes, 2018

Alan Stern and David Grinspoon, *Chasing New Horizons: Inside the Epic First Mission to Pluto*, St. Martin's Press, 2018